The Brain and Conscious Unity

Petr Bob

The Brain and Conscious Unity

Freud's Omega

Springer

Petr Bob
Center for Neuropsychiatric Research of
Traumatic Stress, First Faculty of Medicine,
Department of Psychiatry
Charles University
Prague
Czech Republic

ISBN 978-1-4939-2699-2 ISBN 978-1-4939-2700-5 (eBook)
DOI 10.1007/978-1-4939-2700-5

Library of Congress Control Number: 2015936299

Springer New York Heidelberg Dordrecht London

Printed on acid-free paper

Springer Science+Business Media New York is part of Springer Science+Business Media
(www.springer.com)

Preface

At the end of the nineteenth century Sigmund Freud proposed his "Project of scientific psychology" and wrote: "The intention is to furnish a psychology that shall be a natural science." Freud started this project with the aim to study mental functions and applied general thermodynamic concepts of energy exchange and its distribution in the mental apparatus. In this context he used the term psychodynamics as an analogy to thermodynamics to emphasize the common principles of all phenomena in nature that as its part involve also mind and its rules. In his 'thermo-dynamics' of mind Freud postulated that neuronal activities tend to move towards states with minimal energy and lower levels of mental 'tension' reflecting intrapsychic 'potential' energy. Based on this postulate mind and brain like any other physical or chemical systems that have 'free' energy tend to move and create structures with lower levels of energy. If in a thermodynamic process the energy is not bound or used as the moving energy, it is transformed into increased disorder in the system characterized by its entropy. The entropy quantifies a level of disorganization and based on this principle also mind and brain may be disorganized as a consequence of lost 'intentional' energy which leads to an increase of the disorder and entropy of the mind manifesting as increased psychological tension, intrapsychic conflict, mental disorganization and other stress manifestations.

In agreement with general thermodynamic principles, mind and brain transform the free energy into mental and behavioral activities and focus it on a target in the process of projection or transference which enables that the free psychic energy is 'bound'. Based on this principle mind and brain tend to prefer dynamic activity patterns related to feelings of balance with lowest possible level of free energy. In this context, Freud suggested that mental integration as a potentiality of the 'balanced' state of the mind without a conflict might be linked to 'neural unity'. In agreement with this basic postulate currently there is evidence that deficits in neural connectivity integrating distributed neural activities are related to perceptual and cognitive states producing disintegration of the mind and 'intrapsychic' conflict. These conflict related activation patterns create stress experiences and produce neural interference, temporal disorganization and deficits in functional integration of distributed neural activities, which in principle might be explained as disturbed 'neural unity'.

But on the other hand, certain internal disunity of the brain is a necessary condition to experience the world because mind senses the outside world through differences of its own states that enable recognition and awareness of the external world and also self-reflective experience of internal processes. This form of internal disunity of the brain and mind likely presents basic code that defines relative differences and enables their recognition in mental and physical space and through this process specific observers may define reality and create observer-specific cognition and experience of the space and time.

Based on these principles space and time observations are mediated through differences among the brain states that enable to process the physical observations. Von Neumann proposed that physical theory for its description of basic laws of the universe requires the existence of a conscious observer whose mental activity is not independent and as a part of this process influences physical measurements. Based on the observer role von Neumann divided the world into three parts. Part one of this division is everything up to the senses of the observer, part two includes the observer's perceptual systems, nerve tracts, and the brain, and part three is the observer's abstract "mind".

From this point of view the abstract mind as a basis for self-reference and integrative mental experience is the same in all observers because all differences may be compared only with respect to unity that creates background and reference for all these changes. But when consciousness as an abstract mind creates background of unity for all possible changes is it possible that it can be 'divided' into plurality of minds or it is the same mind in everyone? Schrödinger called this problem "arithmetical paradox" representing antinomies of dividedness and unity of the conscious egos and thought: "but, inconceivable as it seems to ordinary reason, you—and all other conscious beings as such—are all in all …".

Mind Crossroads

Lot of things and dreams and many open doors,
many ways and promising scenes, beautiful links and hopes.
No one sees which hopes and dreams are vain.
Which way is the right one, and which is only pain.
And how to sustain it and be able to see,
what lights in the soul are the gateway to the sea,
where all hearts and souls like one seem to be.

There between lands in space of soul, where to move means to dance.
In a stream of lights that is not possible to see by eyes,
in invisible time, I met your mind.
There future is earlier than past and presence is not existence, to love is not desire, and freedom is not in time as a sequence of events.
In this mysterious place I met your heart,
where in front of me is the same like behind.
There I found flowers of soul in freedom to know,
in deepest space of love.

Content

Chapter 1
Psychodynamics as 'Thermodynamics' of Mind

In 1895 Sigmund Freud proposed his "Project for a scientific psychology" where he proposed psycho-dynamics as a discipline based on thermodynamic principles. Freud focused his attention on the 'energy' and its distribution in the mental apparatus and outlined how psychopathology may be related to normal psychology. Within this 'thermo-dynamics' of the mind he proposed basic terms of 'psychodynamics'.

Based on the physical principle of 'inertia' Freud postulated that a level of neuronal energy tends to move towards states with minimal energy, similarly like in any other physical or chemical systems, as a consequence of the second law of thermodynamics. According to this postulate and analogically to thermodynamics also mental energy tends to minimize itself as a response to mental 'tension' reflecting intrapsychic 'potential' energy. In this context, Freud outlined the concept of 'bound' energy and 'free', or 'unbound', (potential) energy related to states of consciousness and the brain. The energy usually tends to transfer into specific organized mind-brain states that, in close analogy to thermodynamics in which energy supplied to heat machines is transformed into mechanical movement and power, enable to generate intentional behavior. Similarly like in the heat machines, where the energy not used by the engine produces 'heat energy' increasing motion related energy of material particles of the engine and cause disorganization, also mind may be disorganized as a consequence of unused and 'lost' energy. In material objects the process of increased disorganization may be for example observed during phase transitions from solid matter to liquid or gas in which crystal structure tends to be disorganized into random or chaotic molecular movements.

According to thermodynamics in general this specific loss of energy generates disarrangement that may be mathematically described as increased entropy characterizing a level of spatial disorder. Following Freud's original contribution, this concept may be applied also for description of the mind that may use its free energy and realize it in the external world but at the same time a part of this 'intentional' energy may be lost and lead to an increase of the disorder in conscious and unconscious processes, and increased entropy of the mind. For example, increased entropy may be observed as disturbed association process caused by stress, disappointment or

P. Bob, *The Brain and Conscious Unity*, DOI 10.1007/978-1-4939-2700-5_1

a traumatic event. In light of several recent findings this original Freud's proposal seems to be promising and may help to integrate current concepts in physics of entropy and information theory with neuroscientific research and may help to find new perspectives for psychological science in the context that Freud in his original project outlined.

1.1 Freud's Project for Scientific Psychology

In 1895 Sigmund Freud proposed his "Project for a scientific psychology" with its main purpose to investigate mental functioning from the quantitative point of view based on thermodynamic concepts similarly like in physics or chemistry (Freud 1895). In this intriguing scientific concept he proposed psycho-dynamics as a discipline based on thermodynamic approach and corresponding principles. Freud focused his attention on the 'energy' and its distribution in the mental apparatus. In the context of energy exchange he outlined how psychopathology may be related to normal psychology. Within this 'thermo-dynamics' of the mind he focused on domains of the unconscious and outlined basic terms and concepts of 'psychodynamics' and psychoanalysis.

Starting point for his work was collaboration with Viena physiologist Ernst Brücke who in 1870s proposed and described the idea that all living organisms are in fact energy-systems similar to non-living systems with respect to thermodynamic laws (Brücke 1873–1874). In this context, Freud applied basic principles of thermodynamics and used the term "psychic energy" and proposed that the human mind is also an energy-system and that psychology description in its investigations may use basic thermodynamic principles (Bernfeld 1949; Jones 1953). The "Project for a Scientific Psychology" (Freud 1895) was originaly a part of correspondence between Freud and Wilhelm Fliess and Freud himself called it "Psychology for Neurologists". The Project was first published in German original in 1950 and its translation into English was published in 1954 (Andersson 1962; Pribram and Gill 1976; Vichyn 2005).

Based on the physical principle of 'inertia' Freud postulated that a level of neuronal energy tends to move towards states with minimal energy, similarly like in any other physical or chemical systems, as a consequence of the second law of thermodynamics. According to this postulate and analogically to thermodynamics also mental energy tends to minimize itself as a response to mental 'tension' reflecting intrapsychic 'potential' energy. In this context, Freud outlined the concept of 'bound' energy and 'free', or 'unbound', (potential) energy related to states of consciousness and the brain.

In principle, this Freud's concept includes two basic viewpoints. The first is related to the neural device and the concept of "neural thermodynamic machine" and its implications for current view of connections between psychoanalysis and neuroscience. The second basic viewpoint implicitly mentioned in the Freud's "thermodynamic" concept represents mathematical and physical implications of the thermodynamic approach for the study of the mind and brain from the point of view of current physics of thermodynamic systems and complexity.

1.1.1 Neural Reductionism in Freud's "Project" and Neuropsychoanalysis

Freud (1895) proposed basic postulates that observable psychological phenomena depend on the neural structures of the brain based on the three systems of neurons phi (ϕ), psi (ψ) and omega (ω) separated by contact barriers that disable free flow of the energy between neurons in the networks. The ϕ-system of neurones is focused on perception of stimuli in the external world and in fact it reflects the world of nature- *physis* (ϕυσιζ). The Ψ-system of neurons is involved in memory and creates basis for the ego structures and the world of *psyche* (Ψυχη). The third system ω-neurons is the last one (and the last letter in the Greek alphabet) and represents ultimate instance of psychological and perceptual reality represented by the two other neuronal systems that might be linked to awareness, self-reflection and self-reference. Based on these three neural systems Freud proposed some explanations of observed psychological phenomena.

Although this division is intuitively reasonable in fact it is a speculative scheme that was not supported by empirical data describing real biological existence of these neural systems and their specific interactions. The division may be understood as substantial neural reductionism of psychological processes including self-reference and conscious experience per se. These neural vehicles phi, psi and omega based on thermodynamic principles are postulated machines producing mind and its basic structures in the psychic apparatus.

Freud (1895) in his "Project" also emphasized one interesting problem, that the psychological unity might be achieved and represented by a neuronal unity. In this context he discussed that psychological disunity arises during creating of a judgement which may cause divergent components that disturb complex thought processes. In fact it implicates that most likely psychological disunity is represented by a certain form of "neural disunity". This issue in principle presents a basic problem for neural sciences how the *psychological disunity* in the *neural system* may be *represented* (Fig. 1.1).

Based on these ideas the Freud's "Project" proposes a basic scheme for the relationship between later proposed psychoanalytical concept and neural sciences with the aim to study relations of the nervous system in close relationship to conscious and unconscious mental phenomena, later described within the framework of psychoanalytic theory. In modern language a basic postulate of this concept is that psychoanalytic explanations and interpretations of mental phenomena refer to processes that may be described using language of neuroscience, physics and other scientific disciplines as originally Freud conceptualized in his "Project" and later based on relationships between growing neuroscience and psychoanalytical thinking other authors also discussed (Beutel et al. 2004; Edelson 1986; Schore

Fig. 1.1 Sigmund Freud (1856–1939)

1997; Kandel 1999; Solms 1997; Westen and Gabbard 2002a, b; Yovell 2000). In this context, Solms and other authors proposed the concept of neuropsychoanalysis (Solms and Saling 1986; Solms 2004; Solms and Turnbull 2011; Panksepp and Solms 2012).

According to this neuro-psychoanalytical concept it is possible to develop substantial links between Freud's neurological and psychoanalytic ideas (Solms and Saling 1986). This neuropsychoanalytical concept is based on biological descriptions of the brain that in principle may be connected with Freud's psychological theories and in clinical practice may provide integration of psychotherapy with psychoactive medications in agreement with current brain imaging data which show that psychotherapy affects the brain in similar ways as psychoactive drugs (Solms 2004). Based on modern concepts, neuropsychoanalysis is aimed to understand the human mind and especially its relationship to personal "subjective experience" the so-called "first-person perspective" that in principle may be linked to processes in the human brain but its existence cannot be reduced only on information processing in neural networks and has specific dynamics that is called 'mind' (Panksepp and Solms 2012). For example, conversion of a stressful event and intrapsychic conflict into somatic symptoms, behavioral disorders or seizures, and on the contrary there is evidence that somatic disturbances influence the mind. In this context, it is possible to suppose that a neural event has its representation on conscious or unconscious mental level and vice versa.

In the current psychoanalytical thinking the "neuro-viewpoint" is a controversial issue and may be misunderstood as neural reductionism that has been repeatedly a subject of psychoanalytic criticism (Pulver 2003; Vivona 2009; Talvitie and Ihanus 2011). In this context some authors discussed limits and clinical insignificance of neuroscience for psychoanalysis "simply because" it is not possible to analyse human experience in neuroscientific terms (Blass and Carmeli 2007, p. 36).

With respect to the concept of causality between mind and brain, it is possible to suppose that both structures in principle reflect each other and are "isomorphic" with respect to general principles of mathematics, in which isomorphism (from Greek isos- equal and morphe- shape) is studied in order to extend insights from one phenomenon to another (Marker 2002; Mazur 2008). In fact, if two objects are isomorphic, then any property described in scientific description of one phenomenon is on a different "isomorphic" level valid for the objects in another descriptive domain. Based on definition of isomorphism this problem is well defined in mathematics and enables to use it for description of scientifically observed natural processes in which the "same" object can be presented to us in different ways. For example "wave-particle dualism" in physics represents typical example of dual-aspect monism in scientific description of physical phenomena, in which the same physical entity, e.g. light may be observed as a wave during interference and as moving particles (photons) in photoelectric effect. On the other hand electron can behave as a particle but in certain conditions also may be a wave and manifest interference, and this principle of duality likely is valid for all material bodies (Buchvald 1989; Penrose 2004) (Fig. 1.2).

Applied to experimental work in science isomorphic description means that if an isomorphism can be found between some fields in which many theorems are already proved and is possible to describe analogies between those fields then they in principle may be linked to a kind of "superstructure" that underlies both of them and represent their conceptual unity in duality. In this context, even when the super-object is presented to us by two different ways and enables two different descriptions, there is a potential description that enables to consider both of them as equal and isomorphic (Marker 2002; Mazur 2008). On this level of general abstraction it is needed to separate a general quality connecting both descriptions and find how this general quality is specifically reduced to create the first one or

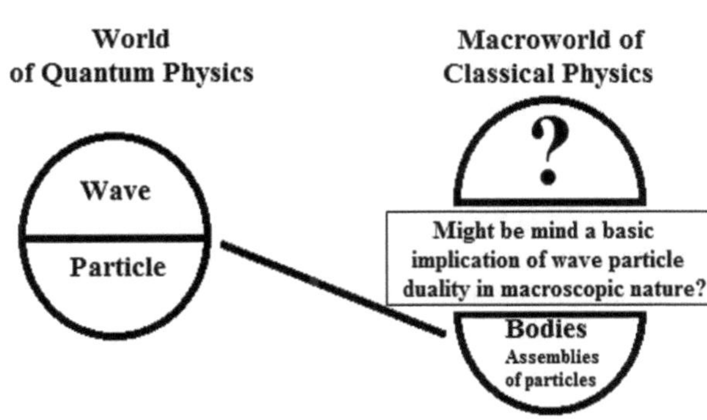

Fig. 1.2 Wave-particle dualism may implicate duality between bodies and minds, in Descartes' concept described as duality of Res Extensa and Res Cogitans

the second descriptive level. In this context brain and mind may be understood as two aspects of the same mathematical (or physical) "superstructure", in the same context in which light and other objects are particles and waves at the same time. In this description brain and mind represent two connected substances that cannot be reduced to each other.

Based on the principle of isomorphism there is a valid assumption that using neuroscience for description of psychoanalytic phenomena may help to study relationships between psychological and neuroscientific viewpoints without reductionism within the concept of dual aspect monism (Talvitie and Ihanus 2011). Talvitie and Ihanus (2011) suggested that this purpose could be possible to achieve within the framework of "neuropsychoanalysis" based on the dual-aspect monism that could be understood as an "interfield" theory providing unifying knowledge connecting both fields and viewpoints. Connections between both fields as related to objectively conceptualized scientific view may be based on "circular causality" principle (Freeman 1999) that explains causal chain of phenomena in the field of mind and in the field of physiological events but also interfield causality which may describe a physiological change due to an event in the mind and vice versa.

In this context, historical foundations of neuropsychoanalysis are closely linked to discussion of the basic philosophical concepts in this field mainly focused on phenomena that could provide scientific description of the brain and mind based on connecting principles the so-called "dual-aspect monism" which may present basic paradigm for new concept of neuropsychoanalysis (Solms and Turnbull 2011).

1.1.2 Psychodynamics as a "Thermodynamics of Mind"

Although the postulated neural systems phi, psi and omega are intuitively of interest in fact they present speculative approach without empirical confirmation. On the other hand using general principles of thermodynamics for understanding brain and mind functions is reasonable and in agreement with basic scientific rules even it is not possible to find how this machine works based on neural principles. Rigorous application of neural thermodynamics that Freud (1895) tried to outline shows that bound psychological energy related to neural excitation may be directly released into motor actions and patterns of behavior representing specific qualities of mind. When this bound psychological energy is linked to conflicting streams of information that disable direct release of excited neural energy this conflict leads to disorganization within the mind. A similar process may be also observed in non-living thermodynamic machines where supplied energy in the usual thermodynamic system is directly released by a movement. When a part of the energy cannot be released by this way, it is lost and produces disorder in the system and its environment that increases entropy. In this context, the "thermodynamic" functioning of the mind and brain suggests that when the regular release of intentional energy is not possible it may lead to increased disorganization in the mind as well as in the brain and increased entropy similarly as in the other thermodynamic systems. This description of increased entropy in the mind may provide basic explanatory model for understanding psychopathology based on thermodynamic principles.

Nevertheless with respect to current findings using thermodynamics as a general description of the mind-brain relationship does not mean that mind can be fully understood using description of its brain, and on the contrary, mind theory may not provide description of the brain phenomena, even if in certain conditions mind may be significantly determined by events in its brain. Both phenomenological levels of description, mind and the brain likely have coupled internal dynamic structures that might be mutually related and causally influenced. In fact it means that (super) structure as a pattern of relations on various levels of organization creates a system organized by these characteristic patterns of its defined components that may be represented on the level of mind and brain and as interactions between them. They create dynamic and metastable structural changes that define dynamics of the system during the time. Based on this principle, brain as well as mind, may be defined on a principle of structural patterns of relations that are created by their basic defined components.

This Freud's original viewpoint suggests that his "Project" in principle is not a reductionistic theory and on a basic level of description includes important theoretical elements that are also contained in modern theories of complexity and self-organizing systems which implicate that increased randomness and disorder in the system lead to different relations of system components and system entropy. In the viewpoint of Freud's proposal it means that the neural excitation may be generated not just externally but also by influence of internally generated energy (i.e. internal drives) that may tend to generate motor activity and patterns of behavior. This internally generated, i.e. self-organizing arrangement and patterning of system components that represents intentional activity may also implicate increased disorder and random-like behavior in the system when the intentional movement cannot be performed, similarly like in any other thermodynamic system. This Freud's proposal in fact means that mind is not governed just by outside environmental influences but also by its internal structure that may cause and influence its activity which in recent science is specific and characteristic quality of the self-organization processes.

For example, Camazine et al. (2003) define self-organization in biological systems as a process in which pattern at the global level of a system emerges solely from numerous interactions among the lower-level components of the system. Moreover, the rules specifying interactions among the system's components are executed using only local information, without reference to the global pattern. The concept of self-organizing systems is in principle in agreement with current neuroscientific evidence that mental disorder may be induced by injurious stimuli from the outside world like chemical substances or various biological or physical influences and also due to internal disorganization related to stress (e.g. Andreasen 1997; Teicher et al. 2003). This increased randomness in the brain may cause relational changes of its components and on the other hand mental disease may be induced by information overload (e.g. Breakspear 2006; Bob 2008). This information overload leading to confusion and increased randomness in mental processing mainly due to painful and traumatic experiences seriously affects mental components and structural patterns of the human mind and produce psychological conflict and mental disintegration as well as physiological and structural brain abnormalities (e.g. Teicher et al. 2006; Bremner et al. 2007).

1.2 Mind and Free Energy

Freud based on the thermodynamic principles proposed that *Id* as a psychological entity producing instinctual drives interacts with the stimuli from the outside world and produces psychological energy which is analogical and in principle the same as energy in any other thermodynamic system (Freud 1923). The psychological energy usually tends to transfer into specific organized mind-brain states that based on thermodynamic principles (similarly like energy supplied to heat machines) is transformed into mechanical movements producing behavioral patterns that enable to generate intentional behavior. Typical form of this intentional activity is an automatic response to a particular stimulus or search related activity when the particular stimulus related to a "wish" is not actually present.

In the case of living "thermodynamic" systems it means that specific innate dispositions and abilities to recognize are linked to internally generated mental and behavioral patterns that are modified and developed by learning. This concept of thermodynamic self-organization is in agreement with principles of ethology as established by Konrad Lorenz. Lorenz in his experimental research with geese and ducks found that certain abilities to recognize are automatically activated without prior learning and developed the concept of imprinting that was in scientific literature for the first time proposed in the 1870s by Douglas Spalding (Lorenz 1935/1970, 1950, 1965). Lorenz described these innate "fixed action patterns" as an underlying predisposition to learn (Lorenz 1950; Fig. 1.3).

The process of innate recognition refers to a specific integration of neural activities that are related to subjective experiences based on functionally distinct cognitive modules (Van De Grind 2002). Later neurobiological evidence has shown that these innate paterns of behaviour are related to specific neural activities in multiple specialized memory systems that define recognition pattern (Freeman 1991, 2000, 2001; Fell et al. 2003; Womelsdorf and Fries 2006, 2007; Jensen et al. 2007; Sauseng and Klimesch 2008; Fries 2009). These specifically different patterns of neural activity enable distinction among various outside stimuli, for example the distinction between memory for song and memory for spatial locations in birds, and similarly it is in humans who have specific memories for various specific behavioral patterns (Sherry and Schacter 1987; Goldsmith 1991; Sherry et al. 1992; Cosmides and Tooby 1995; Shettleworth 2000, 2009). In this ethological concept behavioral states are described as essential components of mental experience that involve specific and unique configurations of a set of physiological, and behavioral states that create psychoneurophysiological patterns of activation which determine discrete patterns of thinking, feeling, and behavioral actions that regularly fluctuate in every individual mind.

Fig. 1.3 Konrad Lorenz (1903–1989)

Particularly important role in the recognition and activation of the behavioral patterns plays attention as a cognitive process of selective concentration on a particular stimulus while other competing stimuli are ignored and inhibited. Selective attention can be defined as a selection among potential conscious contents and a specific function of attentional mechanisms is to bring different events to consciousness (Baars 1988, 2002). This process is related to global distribution of information that is located in brain regions underlying conscious processing (Baars 1997, 1999, 2002). During recognition and selective attention processing, these dynamical interactions of various independent activities produce neural "recognition patterns" that enable to compare sensory information with patterns in memory and distinguish differences between these patterns of possible interpretations of the received information during selective attentional processing (Desimone and Duncan 1995; Baars 1988, 2002; Kanwisher 2001). The event of attentional awareness is related to gestalt perception and also memory processing directly related to a stimulus (or endogenously generated search related activity) accompanied by increased synchrony between sensory and prefrontal regions (Fell et al. 2003; Kaiser and Lutzenberger 2003; Womelsdorf and Fries 2007; Sauseng and Klimesch 2008; Fries 2009).

Specific process that plays a significant role in the recognition and perception presents an event of conflicting streams of information that disables immediate flow of psychological energy into behavioral patterns. Freud also mentioned this conflicting process in his "Project" and characterized it as psychological disunity that arises during the creating of a judgement leading to divergent components that disturb complex thought processes. When the recognition and interpretation are not possible because of a cognitive conflict then the free energy cannot be transformed into a configuration of physiological and behavioral states and create psychoneurophysiological patterns of activation and behavior. Freud described this process in psychological terms as repression of the libido energy that creates secondary consciousness, intrapsychic conflict and mind disintegration (Breuer and Freud 1895; Ellenberger 1970). In thermodynamical terms this increased disorganization due to a cognitive conflict in the system in fact corresponds to increased entropy in the system and loss of instinctual and intentional energy.

1.3 Entropy of Mind

According to thermodynamics in general this specific loss of energy that generates disarrangement may be mathematically described as increased entropy characterizing a level of spatial disorder. Following Freud's original contribution, this concept may be applied also for description of the mind that may use its free energy and realize it in the external world but at the same time a part of this 'intentional' energy may be lost and lead to an increase of the disorder in conscious and unconscious processes, and increased entropy of the mind. For example, increased entropy may be observed as disturbed association process caused by stress, disappointment or a traumatic event. In the light of several recent findings this original Freud's proposal in his "Project" seems to be useful and may help to integrate current concepts in physics of entropy and information theory with neuroscientific research and help to find new perspectives for psychological science.

With respect to current findings there is a basic dilemma whether structural "disorder" related to increased entropy in the mind and biological system represents only increased randomness in the system or it is also related to specific and different levels of organization that tend to "compensate" actual or potential increase of entropy in the system. In this context, a response to various injurious stimuli that specifically may emerge as mental or physical disease may be linked to increased randomness and entropy or it is possible that the response is not random and present "meaningful response" of the self-organizing system to these stimuli. This problem seems to be critical for psychoanalytical viewpoint because when a mental disorder is only a transition increasing randomness in the mind then this pathological process is psychologically and biologically determined. On the other hand mental diseases may be linked to re-organization and new self-organization of systems components that serve as compensatory defense.

Historically this dilemma is contained also in basic principles postulated by Freud. Based on his scientific and therapeutic experience Freud proposed that even if mental disorder may be seen as an increased randomness of mental processing, in fact this randomness is caused by intrapsychic conflict, which determines that some mental aggregates that are integrated in healthy state of mind may be pathologically disintegrated which is usually linked to psychological defense mechanisms.

1.3.1 Thermo-Dynamics and Brain Self-Organization

Similarly like in the heat machines, where the energy not used by the engine produces 'heat energy' and entropy leading to increased disorganization, also mind may be disorganized as a consequence of unused and 'lost' energy. In material objects the process of increased disorganization may be for example observed during phase transitions from solid matter to liquid or gas in which structure tends to be disorganized into random or chaotic molecular movements (Glansdorff and Prigogine 1971; Fig. 1.4).

Basic findings and theoretical concept of "thermodynamic" processes during transformation of different forms of energy developed Carnot in 1824. Carnot discovered that in principle any engine cannot produce more energy than is supplied for its functioning (the so-called first law of thermodynamics which is an expression of the conservation of energy). This principle states that energy can be transformed from one form to another but cannot be created. In addition, he found that energy input provided for engine functioning must be higher than the output, which means that its effectivity=output/input is lower than 1 (the so-called second law of thermodynamics). This principle implicates that any energy transformation in engines, physical systems, chemical processes and also in biological systems leads to an energy loss on the output and called this phenomenon "entropy". Carnot also provided mathematical formulation of this principle that enables to calculate the energy loss during transformation (Perrot 1998; Haynie 2001; Feynman et al. 2005; Fig. 1.5).

Crucial solution of this problem proposed Ludwig Boltzmann in 1890s, who for the first time formulated the so-called statistical interpretation of the entropy. He found that the heat "energy" generated by macroscopic physical, chemical,

Fig. 1.4 Nicolas Léonard Sadi Carnot (1796–1832)

Fig. 1.5 Ludwig Eduard Boltzmann (1844–1906)

and biological processes, increases disorder on microlevel. In his statistical interpretation Boltzmann proposed that the entropy measures the disorder in an ensemble of microparticles that forms the body, molecules, atoms, etc. [quantitative expression of the entropy (S) is: $S = k \log W$, where k is the Boltzmann constant ($k = 1.38065 \times 10-23$ Joule/Kelvin; log is natural logarithm and W is a quantitative measure of the disorder in a very large ensemble of the microparticles] (e.g. Schrödinger 1944).

This basic principle on general level likely is valid for all energy systems and Freud following Brücke's contribution in 1870s (Brücke 1873–1874) proposed and described the idea that the basic thermodynamic principles may be applied also for description of the brain and mind. At this time Boltzman's interpretation of the entropy as a cause of disorganization was not known and the idea that unused energy and increased entropy lead to disorganization therefore was not included Freud's theory. Later research developed these findings and used the concept of entropy as a basic term for the study of brain functions in various studies (e.g. Glassman and Malamut 1977; Germine 1993; Friston et al. 1995; Friston 1997; De Lucia et al. 2005; Stam 2005; Steuer et al. 2004; Bob 2007) where the changes leading to organization or disorganization of the system's components are closely linked to the term "complexity" which basically describes phenomena that emerge from collections of interacting objects (Johnson 2007). Therefore in principle, the processes reflecting levels of complexity that manifests as organization or disorganization in the mind and brain can be described using entropy reflecting order or disorder of mental components and interacting parts in the brain.

1.3.1.1 Brain and Complexity

In principle the brain functions enable encoding information and connectivity of various aspects of brain information that create information clusters and patterns of neural activity that represent mental contents in the mind. Based on the thermodynamic approach the energy related to these patterns of neural activation tends to be discharged in behavioral patterns. When discharge of the energy is not possible this process leads to repression of the libido energy, intrapsychic conflict and mind disintegration (Breuer and Freud 1895; Ellenberger 1970). In thermodynamical terms this cognitive conflict and repression lead to increased disorganization and increased entropy in the system and loss of instinctual and intentional energy.

On the other hand Freud (1930) discussed the term "sublimation" as a process that enables to redirect psychological energy from the repressed "wishes" related to specific mental contents, neural activities and patterns of behavior to other meaningful forms. This novel form of brain and mental activity related to sublimation in principle likely is linked to rearrangement of brain neural patterns that partially or sufficiently compensate disorganization related to repression and increased entropy.

Particularly important aspect of repression is subjectively experienced loss of meaning that is linked to intentional activity which implicates that using thermodynamic concept means that increased entropy leads to disorganization in the mind and loss of meaning and intentional activity. On the other hand sublimation means that the intentional energy may be used in a different meaningful way, which with respect to repression decreases entropy of the mind and increases its organization, meaningful order and creates novel ideas (Fig. 1.6).

Similar description from the point of view of physics and biology proposed Erwin Schrödinger in his famous book "What is life" (Schrödinger 1944). Schrödinger thought that generate meaning and information means to produce order and "negative" entropy, which suggests that increased statistical entropy means loss of

Fig. 1.6 Erwin Schrödinger (1887–1961)

Fig. 1.7 Claude Elwood Shannon (1916–2001)

information. In addition later scientific evidence indicates that the process of information generation is a spontaneous tendency to create dynamical order that emerges in a form of coherent patterns on various levels from molecules to behavior that is denoted as self-organization (Ashby 1947; Prigogine and Stengers 1984; Hess and Mikhailov 1994; Haynie 2008; Fig. 1.7).

Approximately at the same time, Shannon in 1940s developed concept of information theory that defines information as an organized sequence of temporal events that create meaning by its reception (Shannon 1948) and organized temporal pattern that creates context (Balian 2003; Scott 2005; Haynie 2008; Volkenstein 2009). In addition, Shannon (1948) proposed the term information entropy as a measure of temporal disorder quantifying loss of information [in bits] that as later findings show produces disorder and increased entropy in the thermodynamic sense.

In this context, information loss is defined as an unreceived input that is presented in time (Scott 2005; Haynie 2008; Volkenstein 2009) or on the other hand the information may be lost due to disturbed ability to recognize. Any information is not possible to recognize without previous intention and context that form pattern for recognition which is compared with the present time situation [when it is possible to say "yes" or "no", i.e. 1 or 0 that defines basic information unit– one bit]. Basic characteristic of the recognition pattern is a clearly defined structure which contains all needed elements for its repeated recogniton (Jain et al. 2000; Duda et al. 2000) and information receipt is related to a time needed for recognition process (Valentine 1989; Dierig 1994; Agre 1997).

In the brain information processing, the loss of information before its reception means loss of the information during signal propagation to sensory cortical areas, or the information loss may be caused also by incomplete information integration or cognitive interference during neural information processing related to conflicting information. Consequently the loss of information caused by incomplete information integration implicates neural disunity caused by spatial separation of several neural activities that process separated "parts" of the information. It implicates that increase in information entropy and information loss may be caused by disintegration of several information channels and processors during its processing, for example due to a conflict that also may increase thermodynamic entropy and cause spatial disorder in the brain that consequently may influence deficits in brain functions.

In summary, information may be lost because of unability to recognize it, due to very short presentation and other factors that disturb perceptual processing, and due to information interference, which on psychological level is represented by intrapsychic conflict.

Freud in his "Project" intuitively used the basic physical thermodynamic laws as general principles that in an implicit form contain a fundamental approach to the mind as a thermodynamic system which in principle is an agreement with later development of information and complexity theory in brain research (e.g. Glassman and Malamut 1977; Germine 1993; Friston et al. 1995; Friston 1997; De Lucia et al. 2005; Stam 2005; Steuer et al. 2004; Bob 2007). In the light of Freud's thoughts about neural unity vs. disunity, current findings suggest that mental disunity is likely specifically presented on a neural level as disrupted organization in neural communication that may lead to loss of information and increase in entropy. This disrupted neural communication likely underlies deficits in mental processing described by various neuroscientific concepts that complementarily reflect interrelated processes between mind and brain underlying disturbances of mental and neural integration.

This concept of mental disunity and its neural representation as brain disunity is in agreement with current evidence on schizophrenia splitting as a an extreme case of mental disintegration that is closely linked to neural disintegration. According to current evidence neural basis of consciousness as an integrative experience likely represents synchronized neural processes that connect distributed brain activities related to various mental events into a coherent whole which is significantly disturbed in schizophrenia (Tononi and Edelman 2000; Peled 1999; Lee et al. 2003). Through these coherent links regulatory functions enable inhibition or enhancement of neural excitability and create representational maps that form a basis for integrative mental processing. This dynamic functional organization likely also enables to create simultaneously active groups of neurons involved in dynamic alterations of communicating neuronal assemblies which may be competitive or may create synchrony and coherent networks. In this context, this ability to come into synchrony (in gamma band or other frequency bands) defines neural integration of the network or its complexity represented by simultaneously active groups of neurons (Tononi and Edelman 2000; Lee et al. 2003; Bob 2012).

Critical role in these integrative processes likely play cortical inhibitory systems that enable modulation of neural plasticity which is manifested by a functional reorganization of synaptic connections (Daskalakis et al. 2007). At the same time the inhibitory systems enable orchestrated activation of a parallel set of inhibitory interneurons that organize cortical processes to an intended action and prevent aberrant activation (Jones 1993). Nevertheless, although disturbances of these regulatory functions are influenced by inhibitory deficits, even normal inhibitory functions may fail because of enhancement of certain signals related to increased neural excitability. In this context, the regulatory functions are vulnerable to both inhibitory failure and signal enhancement which suggest that disruptions of temporal binding in distributed neural networks may be influenced by multiple mechanisms. Basic role in these regulatory functions likely play various neuromodulatory processes that have a profound influence on the cerebral functions through their effects on the neural excitability and synaptic functions that are significantly disturbed in schizophrenia and other psychiatric disorders (Gordon and Hen 2004; Gray and Roth 2007; Kruglikov and Rudy 2008).

According to recent evidence, the process of disturbed neural integration leading to increased or decreased functional segregation among groups of neurons may be quantified using concepts from statistical information theory and in particular by defining a measure of neural complexity that could provide a possible explanation for a failure of stability and self-regulation related to disorganized cognition in schizophrenia (Bob 2008; Breakspear 2006; Sporns et al. 2000; Sporns et al. 2002; Bob et al. 2009; Balduzzi and Tononi 2008; Tononi 2004; Tononi and Koch 2008). An increase in complexity is often associated with symmetry breaking and the ability of a system to have different states, which is also associated with a decrease in coherence in space over the long range (Weng et al. 1999). The mathematical and physical concept of neural complexity is characterized by dynamic changes between integration, related to functional connectivity, and segregation, linked to functional specialization of distinct neural subsystems (Tononi et al. 1994; Bob 2012).

According to recent findings various mutual relations in the brain are mediated by numerous and rapid changes in brain complexity (Elbert et al. 1994; Friston et al. 1995; Tononi et al. 1996, 1998; Tononi and Edelman 1998; Sporns et al. 2000; Freeman 1991, 2000, 2001). These changes in complexity are represented by various levels of dependence and/or independence among oscillating neural assemblies with characteristic transient periods of synchronization and integration with lower complexity. There is evidence that these transient synchronized oscillations dynamically linking neurons into assemblies at gamma frequencies (30–100 Hz) are closely associated with sensory processing, attentional selection, effective sensory-motor integration and also play an important role in working and long-term memory (Fell et al. 2003; Womelsdorf and Fries 2006, 2007; Jensen et al. 2007; Sauseng and Klimesch 2008; Fries 2009).

Recent evidence indicates that brain functions related to consciousness and attention require multiregional functional interaction and large scale integration of multiple neural assemblies (Varela et al. 2001; Fries. 2005). According to several findings the interacting neuronal assemblies represent basic functional units in brain information processing that may behave independently with lower level of integration among the units or may be more dynamically integrated into large subsets of neurons that behave coherently with synchronous activity (Seth et al. 2006; Edelman 2003; Elbert at al. 1994; Lutzenberger et al. 1995; Molle et al. 1997; Stam 2005). This number of complex interlinked and simultaneously active neural states is reflected in neural complexity that is mathematically defined as a number of independent variables that determine behavior of the system (Weaver 1948).

Detailed studies of mechanism of consciousness and complexity reported Tononi and Edelman (1998) and in later studies also Sporns et al. (2000, 2002). In contrast to conventional approaches to understanding consciousness that are generally concerned with the contribution of specific brain areas or groups of neurons Tononi and Edelman (1998) tend to find what kinds of neural processes present key properties of conscious experience. They applied measures of neural integration and complexity and proposed the dynamic core hypothesis about the properties of the neural substrate of consciousness. In agreement with usual evidence, Tononi and Edelman (1998) postulated that conscious experience is integrated (i.e. each conscious scene is unified) and also highly differentiated by a huge number of different experienced states of consciousness and provided tools for measurement of integration linked to functional clustering and differentiation representing neural complexity which is possible to apply on actual neural processes.

In this context, concept of complexity can be applied to neurophysiological data and enable to evaluate the degree to which neural processes are integrated and/or differentiated (Friston et al. 1995). It is also possible to compare the values of neural complexity in different cognitive and arousal states or empirically tests the relationships between brain complexity and levels of conscious experience (Tononi and Edelman 1998). Because consciousness is related to high level of functional interaction among neurons, it is possible to predict that the complexity of the dynamic core could correlate with the conscious state of the subject (Tononi and Edelman 1998).

According to recent findings there is evidence that neural complexity may be assessed from EEG records. In this context several studies reported relationship between attentional functions and brain EEG complexity reflecting information processing. For example, Lutzenberger et al. (1995) reported that increased EEG complexity indicates an increase in simultaneously activated neural assemblies. Further studies also reported that EEG complexity was substantially higher during imagery than during actual sensory stimulation (Birbaumer et al. 1993; Schupp et al. 1994; Molle et al. 1997) or experimental cognitive load (Bizas et al. 1999; Meyer-Lindenberg 1996; Micheloyannis et al. 1998, 2002; Mölle et al. 1995, 1997; Stam et al. 1996; Tomberg 1999). Further studies also indicate that complexity is significantly lower during full alertness than during drowsiness (Matousek et al. 1995) and similarly also clear alertness during a state of meditation has shown to be associated with a decrease in EEG complexity (Aftanas and Golocheikine 2002). Consistently with these data it has been reported that divergent creative thought is associated with higher EEG complexity whereas convergent analytical thought was related to the lower complexity (Molle et al. 1996). In this context also other studies show that neural EEG complexity reflects the attentional mode related to processing of cortical stimuli (Pritchard and Duke 1995; Molle et al. 1995, 1996, 1997). Taken together these results suggest that attentional narrowing decreases complexity and causes a reduction in neural competition in connection with an inhibition of neural assemblies irrelevant for task completion during selective attention (Lutzenberger et al. 1992).

According to current findings increased neural complexity observed from the EEG and other psychophysiological measures reflects processes during activity of independent areas that enable fast parallel information processing that runs in a distributed mode (Klonowski et al. 1999; Sammer 1996; Elbert et al. 1992; Svetlak et al. 2009; Bob et al. 2009, 2010). This means that numerous processes from sensory and cognitive channels are executed simultaneously and this desynchronized neural state may be related to active information processing in the cortex (Tirsch et al. 2004).

Tononi and Edelman (1998) proposed that consciousness is not a thing, but a process or changing stream on a time scale of fractions of seconds. In agreement with William James (1890) they emphasized that a fundamental aspect of consciousness is an integrated state although at the same time there is evidence that distributed neural activity, particularly in the thalamocortical system, is essential for conscious experience (Edelman 1989; Picton and Stuss 1994; Newman 1995). Therefore it is possible to suppose that interactions among neuronal assemblies in distributed brain areas create a unified neural process corresponding to a multimodal conscious scene (Edelman 1989). In this context, Tononi and Edelman (1998) suggested that a key neural mechanism underlying conscious experience is represented by interactions between posterior thalamocortical areas that are involved in perceptual categorization and anterior thalamocortical areas related to memory processing and action planning (Fig. 1.8). Tononi and Edelman (1998) suggested that such rapid interactions are achieved through the process of reentry, i.e. the ongoing, recursive, highly parallel signaling within and among brain areas. They used large-scale computer simulations and have shown that reentry can achieve dynamical integration of distributed and functionally specialized neuronal groups in a unified neural process (Tononi et al. 1992; Lumer et al. 1997; Tononi and Edelman 1998). In agreement with this current evidence also changes in conscious experience driven by external stimuli, memories, mental images or dreams are related to changes in activities or deactivations of specific widely distributed brain areas (Roland 1993; Frackowiak 1997; Tononi and Edelman 1998).

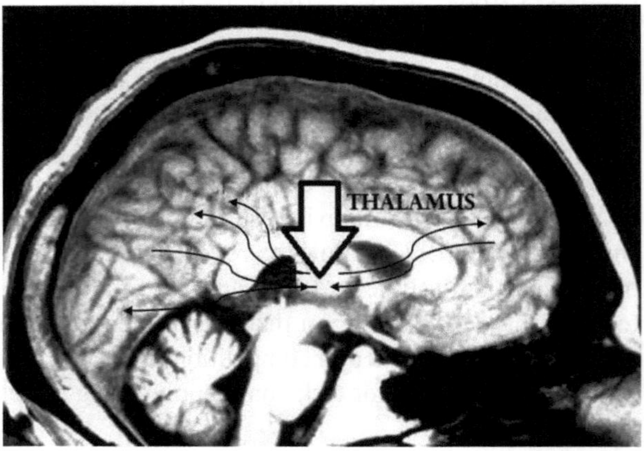

Fig. 1.8 Processes underlying conscious experience are represented by interactions between posterior thalamocortical areas that are involved in perceptual categorization and anterior thalamocortical areas related to memory processing and action planning

Some modeling studies suggest that a specific sign of effective reentrant interactions are short-term temporal correlations between involved neuronal groups (Tononi and Edelman 1998). Other studies also indicate that various kinds of cognitive tasks are related to short-term temporal correlations among distributed populations of neurons in the thalamocortical system (Bressler 1995; Tononi and Edelman 1998). In addition, reported magnetoencephalographic study of binocular rivalry which is simultaneous constant presentation of two different images into each eye indicates that awareness of a stimulus, i.e. subject sees one of them and then after shift the other, is related to increased coherence among distant brain regions (Tononi et al. 1998a). This condition of fast, strong, and distributed neural interactions may explain why feeble, degraded, or short-lasting stimuli often are not consciously perceived, even though they may produce a behavioral response such as perception without awareness (Marcel 1983; Merikle et al. 2001).

Tononi and Edelman (1998) applied methods of functional clustering and found that a subset of distributed elements within a system gives rise to a single, integrated process in cases when these elements interact significantly more strongly among themselves than with the rest parts of the system. This interaction means that they form a functional cluster which is possible to measure by mutual information as a level of integration (Tononi et al. 1998b; Papoulis 1991; Tononi and Edelman 1998) [i.e. when the level of integration calculated among all neurons within the subsystem is higher than level of integration that the same neurons of this subsystem have with neurons out of the subsystem then the subsystem presents the functional cluster]. For example, it is possible to compare levels of synchronous firing among cortical regions and between cortex and thalamus (Tononi et al. 1992; Lumer et al. 1997; Tononi and Edelman 1998). Functional clustering also enables to define system complexity as a number of its parts (i.e. number of clusters) that have higher level of integration within the subsystem than is a level of integration that neurons of this subsystem have with neurons out of the subsystem. In this context, high values of complexity reflect the coexistence of a high degree of functional integration which form "dynamic core" within a system (Tononi and Edelman 1998).

In principle the process of detecting functional integration of specific complex neural patterns creating an unified whole enables recognition and conscious experience of an object defined by its components during selective attention processing. These dynamical interactions of various independent activities produce neural "recognition patterns" that enable to compare sensory information with patterns in memory and distinguish differences between these patterns. This process likely is based on repeated comparisons of various competitive neural patterns representing possible interpretations of the received information during selective attentional processing (Desimone and Duncan 1995; Baars 1988, 2002; Kanwisher 2001). Selected interpretation from this competition and its neural pattern represents the output of the recognition process, which is subjectively experienced as mental representation of the interpretation.

When this recognition process fails and disables to select an interpretation from this competition and its neural pattern it may lead to cognitive interference (for example during Stroop task: "What is color of this word- WHITE "). In this con-

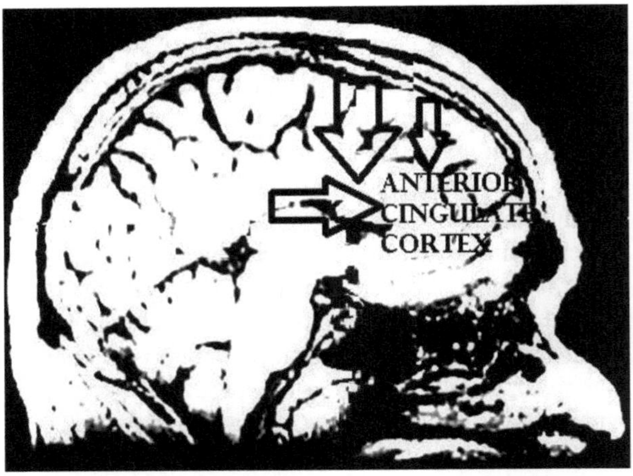

Fig. 1.9 Selection of an interpretation during selective attention and conflict monitoring is closely related to higher activation of anterior cingulate cortex (ACC)

text the cognitive conflict is caused by failed discrimination among mental events and conflict monitoring is closely related to higher activation of anterior cingulate cortex (Bunge et al. 2001; Paus 2001; Baars 2002). Anterior cingulate cortex (Fig. 1.9) is a part of the central network that includes also insula, amygdala, and hippocampus and other medial temporal lobe structures that integrate emotional and cognitive information and have a modulatory role on lower brain centers that control autonomic nervous system and modulates autonomic responses (Benarroch 1993; Critchley 2002).

By this way lasting psychological conflict, for example in schizophrenia could be linked to disruptions in limbic system network integrity and particularly the cingulate gyrus that play a significant role in the pathophysiology and psychopathology of schizophrenia as suggested by brain imaging studies (Fujiwara et al. 2007).

In principle, Freud's (1895) thoughts that neural disunity is caused by intrapsychic conflict, which determines that some mental associations may be splitted (or disintegrated) and lead to neural disintegration, reflect basic concept of complexity in neural sciences. Based on this concept it is likely that the neural complexity may reflect a number of independent neural assemblies and a level of their unity or disunity. In this context, heightened complexity in principle might be related to a number of independent clusters in associative chains or ideas due to disorganization in cognition and mental associations. In this context, it is possible to suppose that complexity could be also related to a number of independent clusters of mental associations that are related to specific emotional states and autonomic arousal (Jung 1907; Weingartner et al. 1977; Stern and Riegel 1970; Bob et al. 2009, 2010). These complex patterns connecting mental and physiological states produce specific patterns of temporal organization or disorganization with increased or decreased functional connectivity that may underlie specific perceptual, emotional and cognitive states (Sporns et al. 2000, 2002).

As a consequence, psychological conflict and increased complexity in verbal associations may lead to increased neural complexity that in the case of schizophrenia therefore might represent a neural process that determines random-like and disorganized cognition, although in fact it may be characterized by complex law- mediated behavior (Leroy et al. 2005). These processes consequently might lead to information overload, deficits in attentional filtering and frontal lobe executive dysfunction (Hotchkiss and Harvey 1990; McGrath 1991; Goldberg and Weinberger 2000). This description is likely in agreement with the view of schizophrenia as producing a failure of attentional modulation, which would be associated with a breakdown in the selective enhancement or inhibition of semantic representations whose underlying networks are widely distributed across left (dominant) temporal and frontal lobes (Nestor et al. 1998, 2001).

Supporting evidence such as word recall studies indicate disturbances in connectivity, linked to associative strength. Furthermore, studies have shown the influence of NMDA receptor antagonists on recurrent inhibition that produces a schizophrenia-like disturbance in association patterns and dysregulation in suppression of associations (Nestor et al. 1998, 2001; McCarley et al. 1999). The dysregulation could be closely related to defective attentional filtering and a failure to inhibit activity of irrelevant neural assemblies (Vaitl et al. 2002; Olypher et al. 2006) and pathologically increased neural complexity (Bob et al. 2009). Such states may implicate chaos that occurs when the neural process involves a large number of complex interlinked and simultaneously active states (Korn and Faure 2003). These conflicting links in associative chains suggest an explanation that the increased complexity (and neural disunity) in schizophrenia could be related to the inner conflict, which is subjectively experienced as "inner chaos" related to increased entropy in the mind.

1.3.1.2 Complexity, Chaos and Self-Organization

Highly organized nonlinear processes related to high system complexity and neural disintegration may lead to a specific sensitive dynamic state of the system called "chaos". In the dynamic state of chaos very small changes in a system can result to very large differences in the system's behavior, the so-called butterfly effect, which means that the flapping of a butterfly's wings in one part of the world later may cause a tornado in a distant part of the world. Chaos represents a dynamic pattern of activity that occurs when the system involves a large number of interlinked and simultaneously active neural assemblies and runs in a desynchronized parallel distributed mode, which can lead to self-organization (Elbert et al. 1994; Freeman 1983, 1991, 2000; Korn and Faure 2003). The concept of dynamical chaos was for the first time developed by French mathematician Henri Poincaré (1854–1912), who studied predictability in a system behavior and found that chaotic pseudo-randomnes is caused by high system sensitivity leading to disproportional changes as a response to stimuli that influence system behavior (Poincaré 1908, 1998; Peterson 1993; Fig. 1.10).

Poincaré worked on problems in celestial mechanics and in his research of the three-body problem discovered a chaotic deterministic system (Folina 1992; Barrow-Green 1997). He found that the motion of a planet in a three-body system can

Fig. 1.10 Henri Poincaré (1854–1912)

be "wild" in its movement and unpredictable and published his novel ideas in 1890s in work "On the Problem of Three Bodies and the Equations of Dynamics". As a consequence of his findings on planetary system he later found general principles indicating that the sensitivity related to very small influences, which significantly decreases ability to predict system's behavior leads to information loss about later system's development. In his "Science and method" (p. 68) Poincaré (1908, 1998) wrote: "A very small, unnoticeable cause can determine a visible very large effect; in this case we claim that this effect is a product of random … However, even if the natural laws were perfectly known, we will ever be able to know the initial conditions with some approximation. If this allows us to know the future with the same approximation that is all we want. We will say that the phenomenon is foreseeable, that it is governed by laws; however this is not always the case, it is possible that very small initial differences lead to very large one in the final state…"

Although the nonlinear mathematical approach to the so-called chaotic phenomena and complexity in nature has its roots in the Poincaré's work in the last years of nineteenth century, its application to the field of psychology and neuroscience is relatively new. A purpose to use this method is the understanding of relatively short periods in the behaviour of a system which are extremely sensitive to very little changes (the so-called sensitivity to initial conditions). This sensitivity during critical times characterizes initiation of new trends in the system's evolution which later may emerge as very different macroscopic patterns of neural activity and mental processes (Elbert et al. 1994; Freeman 1983, 1991, 2000; Birbaumer et al. 1995; Kantz and Schreiber 1997; Meyer-Lindenberg et al. 2002; van Putten and Stam 2001; Faure and Korn 2001; Globus and Arpaia 1994; Korn and Faure 2003).

Several authors proposed that chaotic transitions may emerge in a wide variety of cognitive phenomena and possibly may be linked to specific changes during development of mental disorders such as depression or schizophrenia (Pediaditakis 1992; Schmid 1991; Barton 1994; Gottschalk et al. 1995; Huber et al. 1999; Melancon and Joanette 2000; Korn and Faure 2003; Paulus and Braff 2003; Bob 2007;

Bob et al. 2009a, b) and might underlie psychological hypersensitivity to outside stimuli and their pathological processing.

Because of this sensitivity and unpredictability these nonlinear dynamical systems, although they may be deterministic, exhibit complex and random-like behavior. As experimental research indicates, values of the measured properties of many biological systems look random and their determinants are frequently unknown because of high complexity of factors that influence state of the living organism (Elbert et al. 1994; Freeman 2000; Dokoumetzidis et al. 2001; Korn and Faure 2003). Concept of randomness relies on evidence that every complex system has a large number of degrees of freedom which cannot be directly observed and are manifested through the system's fluctuations (Elbert et al. 1994; Dokoumetzidis et al. 2001; Freeman 1991, 2000, 2001). Recent research shows that the chaotic deterministic dynamical systems display the random-like behavior often indistinguishable from truly random processes (Elbert et al. 1994; Dokoumetzidis et al. 2001). But there is evidence that chaotic dynamics tends to produce a spontaneous order and patterns of organization– self-organization (Elbert et al. 1994; Freeman 2001; Dokoumetzidis et al. 2001; Korn and Faure 2003). The self-organization patterns typically are linked to instability states that may result to new modes of behavior. The sudden phase transitions called bifurcations present a typical form of system's behavior which is deterministic and characterized by typical modes of behavior called the "attractor". The attractor represents typical states of the system that describe its behavior which is not random and therefore it is compressed to a limited subset of all possible states in the "state space" [state space = all possible states that a system in principle could have] (Elbert et al. 1994; Freeman 2001; Dokoumetzidis et al. 2001). On the other hand, random system has not any restriction to its behavior ("it is random") and there is not limitation for its behavior in the "state space"

In this sense, it is possible to use the term state space for various phenomena from position of a particle to states of the human mind (in many cases may be useful to perform graphical representation of the state space). Deterministic systems are in their behavior strictly limited and resulting behavior such as movement of a very small body (for example in gravitational field) is exactly defined and predictable, which means that in constant condition (of gravitation) the body falls and then does not move. Behavior of such a system is strictly limited and defined in the time and space (which together defines the state space of the body). In this context, state space may be visualized by state space diagram in which every possible state of the system corresponds to a unique point in the space and the number of dimensions or parameters of this space represents degree of freedom of the system and every dimension may be represented as an axis. Of course in cases of complex behavior it is difficult or impossible to imagine the state space, similarly like we cannot imagine cube or sphere which has more than three dimensions. In this context, the term multidimensional space represents analogy with usual experience which is used for definition of mathematical terms. For example, the state space (that includes spatial and time dimension) of the mechanical system may be described by all possible values of position and momentum or in the thermodynamics states or phases of a chemical system which may be described as function of pressure, temperature or composition (Elbert et al. 1994; Dokoumetzidis et al. 2001).

These descriptions of complex behavior using chaos theory are also related to non-equilibrium thermodynamics that described nonlinear chemical oscillating systems in which chemical reactions are not governed by equilibrium thermodynamic behavior and may remain for a very long time (Fox 1990; Epstein and Pojman 1998; Ott 2002). In this context, chemical oscillating systems provide chemical models also for nonequilibrium biological phenomena (Nicolis and Prigogine 1977; Harrison 1993; Erdi 2007), for example, the so-called Belousov–Zhabotinsky reaction (Tyson 1979; Zaikin and Zhabotinsky 1970). Based on these principles Alan Turing in 1952 published his study on "The Chemical Basis of Morphogenesis". In his work Turing (1952) proposed how chemical substances, called morphogens, might react together and create main phenomena of morphogenesis and discussed chemical inequlibrium dynamics that genes may use to create morphological structures.

Specific aspect of dynamical and chaotic systems is their limited behavior and limited space of occurrence in the state space, similarly as other deterministic systems, but their behavior has limited predictability or it is unpredictable in the space and time. Specific form of behavior of the chaotic system defined by an attractor includes spatial and time dimension of all its possible states in the past and future that can be described as a "geometrical object" in the state space. In another words this means that the dynamic and chaotic systems are neither deterministic nor random. Scientific description of complex macrosystem such as living organism may be defined by various complex "state functions" such as temperature, blood pressure, blood flow or electrical activity such as EEG, ECG and electrodermal activity (EDA), and also other physiological, behavioral or cognitive characteristics (Freeman 1991, 2000; Elbert et al. 1994; Globus and Arpaia 1994; Gottschalk et al. 1995; Huber et al. 1999; Melancon and Joanette 2000; Faure and Korn 2001; Meyer-Lindenberg et al. 2002; Korn and Faure 2003; Paulus and Braff 2003; Breakspear 2006; Bob 2007; Bob et al. 2009a, b).

Seminal contributions to this field of research reported Walter Freeman, who was particularly interested to explore how brain generates cognitive processing, intentionality and meaning. His main body of research has been focused on EEG study of perceptual processing in rabbits. In his research Freeman found that activity in the olfactory cortex is chaotic, and proposed that chaos could underly basic forms of collective neural activity in perceptual processing including ability to access memorized sensory patterns and learning novel sensory information (Freeman 1991, 2000, 2001; Skarda and Freeman 1987). Freeman also proposed that chaos may explain brain ability to respond flexibly to the outside world and to generate novel activity patterns that are subjectively experienced as "novel" ideas, generated by unpredictable attractors that enable complex dynamic behavior of the brain and intentional behavior (Freeman 1991, 2000, 2001; Skarda and Freeman 1987). In this context, chaos theory enables to understand collective neural activity and brain functions as a global integrative process based on dynamic collections of attractors which form an "attractor landscape" generated in the web of synaptic connections and representing behavioral "intentional" patterns that may be modified by learning (Skarda and Freeman 1987; Freeman 2000).

Within this framework Freeman (1999) proposed that the linear view of the stimulus-response reflex determinism is not appropriate concept for the behavioral dynamics and suggested that it is needed to study behavioral responses and intentional behavior as consequences of nonlinear chains of various stimuli and responses. Freeman (1999, 2000) also suggested that chaotic and complex self-organization of multilevel interactions between microscopic neurons in assemblies and the macroscopic emergent states is not possible within the concept of "linear causality", and must be replaced by "circular causality" (or reciprocal causality) that enables to reflect extensive relations of mutual dependencies, actions and influences. Although neurophysiological basis of these integrative processes is only partially understood, the concept of circular causality as a formal semantic description of brain dynamics related to chaotic self-organization and multimodal macroscopic patterns of neural activations may help to explain some functions of consciousness and intentional actions.

Taken together, the brain as a physical system has a unique ability to combine neural bits of information into complex patterns of activities that enable mental functions and behavior. This ability of the brain self-organization to coordinate neural microstates and "bind" them into complex patterns of information enables to use and extrapolate basic terms used in thermodynamics and information theory into specific brain information processing. In the brain ordered states are based on coordinated neural activities related to information processing based on series of timely and spatially organized microstates (the so-called neural binding).

In this context, interfield theory represents theoretical postulate suggesting that connections between the field of mind and brain may be build on scientific and experimentally testable principles. These principles most likely are linked to isomorphism based on mutually corresponding levels of order or disorder in the brain as well as in the mind that enables measurable distinguishing between well-ordered or disordered states of the mind and the body. Similarly, within the framework of the dual aspect monism it is possible to suppose that conflict and splitting of the mind are represented as neural disunity due to information disintegration that manifests in the brain as well as in the mind.

In accordance with these principles of general thermodynamics and neuropsychology, Freud (1895) proposed that mind and brain transform the free energy into intentional mental and behavioral activities and suggested that mental integration as a potentiality of the "balanced" state of the mind without a conflict might be linked to "neural unity".

1.3.2 Brain Information Theory and Entropy of Mind

Basic principles to connect information processes in the mind with the physical world of the brain indicate findings that loss of information increases disorder in the neural system. This relationship is based on connecting principles that enable to find relationship between Shannon's concept of information and thermodynamical entropy explained by Boltzmann's statistical interpretation, (Scott 2005; Volkenstein 2009). Interesting and useful concept for description of these processes in mathematical language proposed Claude Shannon with his formulation of the term information entropy (Shannon 1948; Shannon and Weaver 1963) and later research has shown that the Shannon's concept of information entropy is compatible with Boltzmann's statistical formulation of the entropy which implicates that the loss of information increases spatial disorder (Scott 2005; Volkenstein 2009). This relationship between thermodynamic and statistical entropy and information entropy is based on the works published by Szilard (1929), Rothstein (1951) and Brillouin (1956), who found the connection between the Boltzmann entropy and Shannon information entropy (Frieden 2004). Szilard (1929, 1964) using a thermodynamic "gedankenexperiment" in the seminal article "On the Decrease of Entropy in a Thermodynamic System by the Intervention of Intelligent Beings" has shown that an observer who performs the experiment, in order to learn, receives information. In this context Szilard also defined today's so-called "Szilard's limit", which means energy that is transformed into increased thermodynamic entropy compensating information receipt, and formulated physical connection between thermodynamic entropy and information (Szilard 1929, 1964; Zurek 1984).

Similar gedanken experiment later published also Brillouin in 1956 (Brillouin 1956; Frieden 2004). In addition, Brillouin (1956) found that information connected with certain specific physical system, which is bound in the system, is related to its entropy. He found that the entropy decreases when information is obtained and the information must be furnished by some external agent whose entropy will increase, i.e. bound information = decrease in entropy = increase in negentropy [i.e negative entropy] and on the other hand information loss = increase in entropy = decrease in negentropy, which means that increase of entropy and loss of information work together (Brillouin 1956).

A significant contribution to the relationship between statistical mechanics and information theory reported also Jaynes (1957), who shown that thermodynamic entropy is proportional to Shannon information needed to define detailed microscopic states of the system. He found that when a system increases its thermodynamic entropy, which consequently means that it increases the number of possible microscopic states of the system, at the same time increases an amount of information needed for description of these additional microstates in agreement with von-Neumann-Shannon expression for entropy as a measure of the amount of uncertainty representing the loss of information.

Simple quantitative formulation of the relationship between entropy and information provides Landauer's principle (Landauer 1961; Berut et al. 2012) which shows that every 1 bit of information is equivalent to kT log 2 of heat [in Joules]. Quantitative expression of the entropy (S) is: $S = k \log W$, where k is the Boltzmann constant ($k = 1.38065 \times 10-23$ Joule/Kelvin; log is natural logarithm and W is a quantitative measure of the disorder in a very large ensemble of the microparticles (Schrödinger 1944; Giles 1964; Lieb and Yngvason 1999). This *relationship be-*

tween information and entropy means that one *bit of information* increases the thermodynamical *entropy* of the world by *k log 2* and on the other hand for example **loss of 1 bit of information in the mind** as an information device **directly increases entropy of the brain** which implicates that *mental disorder as a loss of information directly implicates increased entropy and spatial disorganization in the brain.*

The relationship between entropy and information has been experimentally verified by Berut et al. (2012), who shown the existence of the Landauer's principle in a model of a one-bit memory. They used a system of a single colloidal particle in a modulated double-well potential and their result demonstrates the link between information theory and thermodynamics. This relationship between information and statistical entropy provides interesting evidence based on fundamental physical laws that the loss of one bit of information in the system increases spatial disorganization of the system related to increase of its entropy (which is numerically equal to kT log 2).

These "information" vs. "entropy" connecting principles have important implications for brain information processing and in principle it means that receiving one or more bits of information decreases a system entropy and deleted bits of information lead to system disorganization. Even likely it is not possible to calculate exactly how many bits of lost information have various psychological conflicts, stress and traumatic experiences, in principle it means that *repression as a loss of information on the conscious level* produces N bits of lost information that may be transformed into mind and body disorganization. This physics of information in principle underlies Freud's and Breuer's *conversion mechanism* in agreement with Freud's concept of mind as a thermodynamic system.

In the light of Freud's "Project" and his later works the crucial term for psychological information loss is the term "repression" representing the withdrawal of psychic energy which may be redirected and compensated by sublimation represented by another contextual system of goals, wishes, thoughts, feelings and emotional contents. In this context, the experienced quality is related to a content of consciousness within the contextual system related to investment of psychic energy (cathexis) representing contextual information related to specific mental and neuronal energy, and to behavioral activities. In the case of divergent conflicting thoughts and emotional activities a basic neural mechanism tends to prevent the brain and mind against "neural" and "information" interference. This defense mechanim prevents psychic unity which leads to executive "repression" of conflicting information in favor of dominant contextual framework due to conflicting information that does not fit into dominant cognitive scheme.

In this context, current neuroscience provide evidence indicating that mental conflict and stress may cause physiological disorganization and structural neurodegenerative changes including decreased volumes of various brain structures (Bremner et al. 1995; Bremner 1999; De Bellis et al. 1999, 2001; Jelicic and Merckecbach 2004; Winter and Irle 2004; Teicher et al. 2006; Bremner et al. 2007; Chap. 4) and brain disintegration corresponds to psychological disintegration mainly due to disturbed synchrony in gamma band and other frequencies (Lee et al. 2010; Peled 1999; Tononi and Edelman 2000; Bob 2011).

On the other hand improved psychological integration due to specific forms of learning for example during psychotherapy likely also influences the brain integrative processes. These findings indicate that learning and memory processes including a wide variety of environmental factors may influence development of synaptic connections through new gene expression and that psychotherapy as a special learning process may specifically influence and modify brain functions, metabolism in specific brain structures and also genetic processes (Kandel 1998, 1999; Gabbard 2000, 2007). Typical and intriguing example how psychological integration may influence brain structures show some studies investigating "mindfulness" meditation (Chiesa, and Serretti 2010; Davidson et al. 2003; Sheridan 2003; Hölzel et al. 2007; Ives-Deliperi et al. 2011). Using this meditation technique direct intriguing influences on cortical thickness and other brain structural changes have been reported (Grant 2013; Hölzel 2011; Luders et al. 2012; Tang et al. 2010, 2012; Lazar et al. 2005; Fig. 1.11).

Recent evidence shows that these neuroscience and psychological findings are in agreement with basic physical principles and specifically mean that disturbed coordination of neural microstates and neural disintegration leading to the information loss implicates increase in statistical disorder. On a neural level the increased entropy and disorder in the system are related to various physical and molecular processes that consequently lead to higher level of disarrangement and disintegration. As a consequence these modern findings in physics and information theory and also recent evidence in psychology and neuroscience do not support *Cartesian dualism* of *mind* and *body*.

Based on this relationship, it is possible to connect time and spatial coordination of neural states with information patterns in the brain and connect brain entropy as a level of disorder reflecting deficits of time and spatial neural integration with various levels and reference frames of the mind and define "entropy" of mind as a

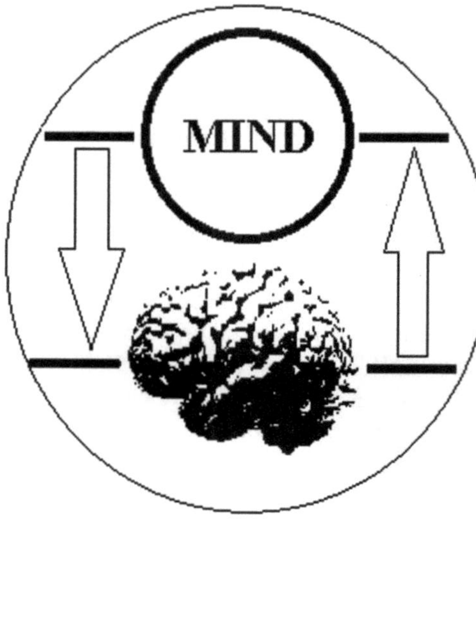

Fig. 1.11 Circular causality between mind and brain

specific level of disorder that is subjectively experienced and reflected in the human mind and behavior.

These physical-information relationships enable to define basic principles of brain-mind information exchange which are based on relationships between brain physical processes and information processing in the human mind. The relationship between statistical (material) entropy in the brain and information entropy related to information content in the human mind therefore represents and constitutes the basic brain-mind information principle, which specifically states that the loss of information during brain information processing implicates increased disorder (and entropy) of the brain and mind.

In this context, isomorphism and interfield view represent valid philosophical concepts that likely are not just general epistemological viewpoints but may be scientifically applied using thermodynamics, complexity and information theory that present connecting basis for the relationship between psychological and neuroscience descriptions that also enable to explain mutual influences between mental and neural level. These principles of isomorphism and interfield view are also in agreement with findings that not only harmful information or damages to the brain lead to psychopathological consequences but also stressful conditions and psychological trauma may lead to brain disorders including structural changes (Bremner et al. 1995; Bremner 1999; De Bellis et al. 1999, 2001; Jelicic and Merckecbach 2004; Winter and Irle 2004; Teicher et al. 2006; Bremner et al. 2007). In this context, also communication patterns related to stressful experiences and specific forms of abuse and neglect may influence pathogical information processing and loss of specific information may cause abnormal brain development and psychological disintegration. From this point of view the principle of information entropy explains recent neuroscience evidence of social stress and deprivation within the framework of general principles of complexity and self-organization indicating that life needs information in its various forms to balance statistical tendency of nature to go from order into disorder.

The brain-mind information entropy is also in agreement with findings documenting the neural disunity and disorder in the brains of schizophrenic patients and there is evidence that spatio-temporal binding and synchronization mainly in the gamma range related to brain information processing is significantly affected in schizophrenia (Tononi and Edelman 2000; Peled 1999; Lee et al. 2003; Ford and Mathalon 2008; Uhlhaas et al. 2008; Uhlhaas and Singer 2010).

According to recent evidence, mental disintegration in schizophrenia is related to disturbed binding and integration of multiple and disparate neural activities underlying cognitive brain functions and consciousness (Singer 1993, 2001; Varela et al. 2001; Fries 2005; Tononi and Edelman 2000; Peled 1999; Lee et al. 2003; Ford et al. 2007). These findings indicate an experimental paradigm for understanding of conscious integration, which suggests conceptualization of schizophrenia as a disorder of neural integration (Tononi and Edelman 2000). The conceptualization of brain and cognitive disintegration in schizophrenia is reminiscent of the concept of 'schizophrenia' proposed by Bleuler in his Dementia praecox or the group of the schizophrenias emphasizing the splitting or disintegration of consciousness

(Bleuler 1911/1955, 1918/1906, 1924). A specific form of mental disintegration in schizophrenia is related to deficits in communication between the frontal and temporal lobes (Ford et al. 2005) related to disintegration of consciousness that probably produces defective self-monitoring and self-experiencing (Feinberg 1978; Feinberg and Guazzelli 1999; Ford et al. 2001, 2007; Poulet and Hedwig 2007; Andreasen 1997; Meares 1999). These communication deficits are mainly caused by disturbed functions of the so-called "corollary discharges" (or efference copy) that represent neural discharges with unique integrative functions. These integrative functions enable monitoring and modification of brain commands themselves before the effector event and enable to inform sensory systems that the stimulation produced by movement is self-generated or produced by an environment, which is crucial for the distinction of self and non-self (Feinberg 1978; Ford et al. 2001, 2005, 2007; Poulet and Hedwig 2007). For example due to a loss of self-reflective information, inner speech may be misidentified as external voices (Ford et al. 2001, 2007; Poulet and Hedwig 2007).

Typical manifestations of mental disintegration in schizophrenia and disturbed order related to information loss were also documented in studies of associative process and other studies focused on discontinuities in schizophrenic thinking. The pseudorandom character of mental processes in schizophrenia was observed in word associations (Jung 1909; Kent and Rosonoff 1910; Moran et al. 1964; Shakow 1980; Goldberg and Weinberger 2000), impaired verbal fluency (Allen et al. 1993; Himelhoch et al. 1996; Vinogradov et al. 2002) and textual analyses of the semantic processing (Manschreck et al. 1979, 1981; Hoffman et al. 1982; Goldberg and Weinberger 2000) indicating deficits in organization of semantic memory in schizophrenia (Davis et al. 1995; Paulsen et al. 1996; Vinogradov et al. 2002).

In summary, these current findings likely provide evidence that extreme cases of mental splitting and disorganization that occur in schizophrenia are linked to increased entropy related to disturbed information processing. A main aspect of neural disorganization is disturbed large scale communication in the brain that represents a form of "neural disunity" is likely related to cognitive conflict and repressed intentionality. In the light of current findings Freud's "Project" focused on thermodynamics of the mind provides intuitive view that the thermodynamic approach is closely linked to intentionality and information, and that repression of this "intentional" energy causes "neural disunity" and increased entropy in the system.

In historical contexts of Freud's time, intentionality had no place in the world of post-Newtonian physics based on (just two) Aristotelian principles "causa efficiens" and "causa materialis" that represent causal principles of (non-living) moving bodies. In this Newtonian and post-Newtonian explanatory scheme that Freud explicitly used as a scientific model for psychology was conceptually necessary to add two other principles that Aristoteles proposed in his "*Metaphysics*", i.e. causa formalis and causa finalis. These principles mean information and intentionality that Freud in the "*Id*" concept presented as "instinctual drives" representing phenomena that distinguish psychological description from otherwise generally valid thermodynamic rules.

Chapter 2
Mind in Conflict and Unity

According to Freud's (1895) psychodynamic concept, the 'free' energy of mind is related to certain unconditioned drives that tend to create certain kinds of thinking, feelings and behavioral patterns that are experienced as 'mental images' and 'internal goals'. The 'free' energy is projected to the outside world through wishes and imagination linked to the process of 'projection' or 'transference' that temporarily identifies—unconscious and implicit—"internal" intentional energy with real events and processes in the external world. The terms of projection or transference are specifically linked to a direction or redirection of feelings and desires to targets in the external world that exist as representations of the internal space of potentialities—goals, wishes and intentions—that have energy that can be bound into these intentions and create mental representational maps.

In close similarity to thermodynamics where free energy can be bound, the process of projection or transference enables that the free psychic energy is 'bound' to a target and transformed into mental and behavioral activities. These include momentarily activities and future plans that create space of meanings in the space and time. In the case that energy is bound to two "mental" objects related to events and objects in the outside worlds experienced as conflicting then it may lead to dissociative state in which mind is splitted into contradictory tendencies. The event related to intrapsychic conflict leads to increased psychological tension, stress and mental disorganization, in which an amount of energy cannot be bound. In this context, the dissociative states are related to mental events that bind free energy into two or more conflicting mental states and behavioral patterns.

Recent ethological and other scientific findings indicate that the human mind contains specific innate patterns of behavior spontaneously generating 'free energy' that tends to be focused and 'bound' to significant and specific mental images, identity, self-representation and neurobehavioral patterns. Based on these innate structural preferences of 'mental association patterns', similarly as in chemical thermodynamics where atoms create preferred forms (or patterns) of 'molecular structures' with lowest levels of energy, also mental structures likely may be created on thermodynamic principles. In this context, by similar way as we study structural

P. Bob, *The Brain and Conscious Unity*, DOI 10.1007/978-1-4939-2700-5_2

determination of various forms in scientific disciplines like physics or chemistry, likely using basic principles of thermodynamics may be possible to find the basic structural processes for description of mental processing.

Based on these general thermodynamic principles there is likely a similar tendency to create dynamic patterns of the mind with lowest possible level of free energy reflecting feelings of 'satisfaction' and balance that likely also define preferred patterns of the brain states. In this context, it is possible to assume that mental integration and unity existing as a mental potentiality related to 'hope' and 'meaning', could represent 'balanced' state of the mind that focuses its free energy to the external world without a conflict, which as Freud suggested in principle might be linked to 'neural unity'.

2.1 Conflict and Energy of the Mind

According to Freud's psychodynamic concept the 'free' energy of mind is related to certain unconditioned —innate or instinctual—drives that tend to create certain kinds of thinking, feelings and behavioral patterns that are experienced as 'mental images' corresponding and representing 'internal goals' and the 'free' energy creates its representation in the outside world through patterns of behavior that temporarily identify the internal energy with real events and processes in the external world.

In this context, recent ethological and other scientific findings indicate that the human mind contains specific innate patterns of behavior spontaneously generating 'free energy' that tends to be focused and 'bound' to significant and specific mental images and "objects" based on innate "fixed action patterns" representing underlying predispositions to learn with relatively stereotyped behavioral elements that are very similar in various populations of animals and humans, similarly as anatomical and physiological properties (Lorenz 1950). These innate paterns of behaviour are related to specific neural activities in multiple specialized memory systems that define recognition patterns (Freeman 1991, 2000, 2001; Fell et al. 2003; Womelsdorf and Fries 2006, 2007; Jensen et al. 2007; Sauseng and Klimesch 2008; Fries 2009). These specifically different patterns of neural activity enable to distinguish among various outside stimuli and create specific memories for various specific behavioral patterns and are modified through conditioned stimuli and learning (Sherry and Schacter 1987; Goldsmith 1991; Sherry et al. 1992; Cosmides and Tooby 1995; Shettleworth 2000, 2009). Current research in comparative psychology also shows that this innate ability of recognize is not linked to just simple patterns of recognition but include also highly abstract forms like numerical cognition (Brannon 2006; Shettleworth 2009). These data provide evidence that human infants and nonhuman primates have innate sophisticated systems representing numbers as language-independent mental recognition of visually presented mathematical objects (Pepperberg 1994; Hauser et al. 1996; Brannon 2006; Shettleworth 2009) that have an ontogenetic origin and a neural basis independent of language (Gallistel and Gelman 1992; Gelman and Butterworth 2005).

Connection of these highly evolutionary developed innate internal patterns with objects in the external world is based on hierarchical scales of less or more general specific qualities and signs contained in processed sensory information that enables recognition. During recognition and selective attention processing, dynamical interactions enable to compare sensory information with information encoded in memory and distinguish differences using neural "recognition patterns". These mental structures enable to distinguish and create relationship between mental intention related to the recognition pattern and its form in the external world. This process likely is based on repeated comparisons of various competitive neural patterns representing possible interpretations of the received information during selective attentional processing (Desimone and Duncan 1995; Baars 1988, 2002; Kanwisher 2001). Selected interpretation from this competition and its neural pattern represent the output of the recognition process, which is subjectively experienced as mental representation of the interpretation.

With respect to the current evidence about attentional processing it is likely that these complex dynamical interconnections forming various neural patterns and also comparisons among them produce "differences" which define basic "codes" for recognition. These differences are likely encoded through specific dependencies or independencies among the neural patterns which define hierarchical scales of identities and differences that are interpreted in the "theatre" of mind as different mutually dependent or independent objects. The brain mechanisms that enable multilevel information processing are regulated within the framework of contextual understanding which strictly determines what details are important for the whole coherent context. This hierarchical processing may provide feedback to lower levels and may initiate reprocessing of any information that enable its reintegration in contexts of various schemes and concepts (Mesulam 1998; Nadel and Jacobs 1998; Lavenex and Amaral 2000). For example, in processes of visual cognition this hierarchical contextual structures have been found in frontal and temporal regions that provide inhibitory feedback to the initial processing in primary visual areas V1 and V2 (Hupe et al. 1998; Bullier 2006; Fabricius 2010).

These findings suggest that the multilevel information processing of perceptual information might be related to executive signals that in the framework of required context determine sensitivity to various details needed for contextual processing and significantly influence processing through various feedback interactions (Gilbert and Wiesel 1990; Francis et al. 1994; Gove et al. 1995; Gray 1999; Grossberg and Grunewald 1997; Ito and Gilbert 1999; Raizada and Grossberg 2001; Kang et al. 2005).

In this sense contextual processing may determine what details of the processed information will actually be presented in conscious awareness and there is evidence that transient synchronized oscillations dynamically linking neurons into assemblies through the process of "binding" mainly at gamma frequencies (30–100 Hz) are closely associated with sensory processing, attentional selection, effective sensory-motor integration and also play an important role in working and long-term memory (Fell et al. 2003; Womelsdorf and Fries 2006, 2007; Jensen et al. 2007; Sauseng and Klimesch 2008; Fries 2009).

The recognition process within various contextual frameworks is determined by hierarchically defined differences that are basically dependent on signal spread to sensory cortices and other structures participating in signal processing that enable to create “neural information patterns”. In this context Shanon’s principle of communication implicates that any event of information propagation and its processing leads to the information loss and produces the information entropy, which defines “uncertainty” of the signal for its receiver. This uncertainty in information processing specifically influences recognition process and increases probability of the recognition error. The information loss produces “uncertainty” or “ambiguity” and negatively influences possibility to recognize. For example, the loss of information during propagation may cause difficulties to distinguish between two similar persons during getting dark and may lead to perceptual ambiguity.

But on the other hand the same effect of uncertainty may be caused by two well perceived conflicting stimuli as for example during the Stroop color-word interference task (is this color “**white**”?) and other experiments related to processing of incongruent and conflicting information (Kanwisher 2001). This relationship between the information loss and possibility to recognize implicates that the main feature of recogniton is clear definition of the recognition pattern which clearly defines the difference between the objects. The “difference” may be significantly decreased or lost due to signal propagation or during perceptual processing leading to conflict.

Recent research of the neural correlates of perceptual consciousness provides extensive evidence from behavioral studies that perceptual information can be represented in the mind/brain also without the subject’s awareness of that information (Crick and Koch 1995; Kanwisher 2001). These findings suggest that perceptual information may be processed and included in various contextual frameworks and that awareness requires access to that information by other parts of the mind/brain (Baars 1988, 2002). The limits on conscious access to perceptual information may not be immutable and conflicting streams of information as well as pathological brain changes may disrupt neural pathways and as a consequence perceptual information represented in one “contextual” neural pattern is not accessed by other parts of the system (Baars 2002; Kanwisher 2001). Experimental conditions that enable to assess modulatory influences on discriminative processes, attentional filtering and contextual processing have been studied also using various methods of hypnotic suggestion and several data indicate that the threshold of consciousness may change during hypnosis. For example, Stross and Shevrin (1962, 1968, 1969) have found alterations of thought contents under hypnosis in investigation of “freely evoked images” after the subliminal presentation. These and other findings suggest that hypnosis leads to heightened access to subliminal stimuli and that thought organization during hypnosis shares some common elements with thought organization during dreaming (Bob 2004). This finding corresponds to similar reported data when subliminally presented images were found in dreams (Fisher 1954; Poetzl 1960). In this context, seems to be likely that various sensory stimuli that has an importance for cognitive processes and adaptive behavioral responses may be influenced by various mechanisms of cognitive modulation and that subliminality may present a relative phenomenon characterized by a sensitivity of discriminative responses (Wortman et al. 1992; Erdelyi 2004a, b; Reingold 2004; Kihlstrom 2004).

According to this evidence even a "strong" and "meaningful" neural representation may not be sufficient for awareness and there is behavioral evidence that perceptual awareness involves not only activation of the relevant perceptual properties but also further construction of an organized representation in which these sensory properties are attributed to their sources in external objects and events (Kanwisher 2001; Baars 1988).

On the other hand there is also evidence that only simple focusing of visual attention on different aspects of an unchanging stimulus has a strong effect on the content and intensity of perceptual awareness (Rees et al. 1999). In this context, numerous studies using single-unit recordings, ERPs and brain imaging have shown modulations of sensory responses by attention, even for a constant stimulus (Desimone and Duncan 1995; Luck and Girelli 1998; Goebel et al. 1998; Kanwisher 2001).

Simple examples of these cases provide ambiguous stimuli leading to alterations of perceptual experience between two different states (Fig. 2.1) such as Necker cube, Rubin's face/vase, moving circle or experiments with binocular rivalry in which different images are projected to each eye (Kanwisher 2001; von Helmholtz 1962). Although the stimulus itself does not change the human observer sees only one of the possible percepts, instead of seeing a mixture of the two possible images. For example, typical experiment with binocular rivalry includes presentation of vertical stripes that are presented to the left eye and horizontal stripes to the right eye. The viewer is likely to see not a superimposition of the two patterns (i.e. a crosshatching plaid pattern), but an alternating sequence in which only vertical stripes will be seen for one moment, and only horizontal stripes the next. Based on research of the binocular rivalry and other competitive phenomena there is evidence that subjective experience alternates in a bistable fashion between being dominated by the one or the second image that results from the interpretation process (Blake et al. 1998; Kanwisher 2001).

Theoretical explanation of the observed phenomena proposed Desimone and Duncan (1995) in the concept of 'interactive competition'. According to this model, competitive interactions across cortical areas result in domination of perceptual representations by properties of a single object. This competition can be biased by either bottom-up factors (e.g. stimulus salience) or top-down factors (e.g. endog-

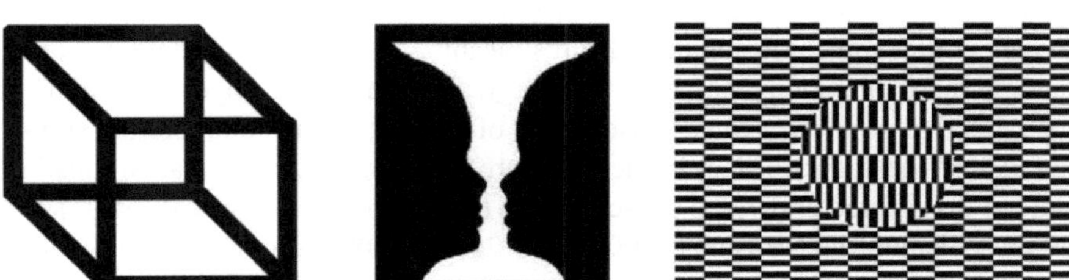

Fig. 2.1 Examples of bisable figures, Necker cube, Rubin's face/vase, moving circle

enous attention). In either case the result is that the various properties of an object represented in distinct cortical regions enhance each other and suppress the representation of competing objects with periods of dominance with mutual exclusivity (Rubin 2003), which results in multi-stability when the two (or more) percepts alternate in a seemingly random manner that likely has nonlinear dynamic features (Aks and Sprott 2003). In this context, attentional processing and awareness are global properties of the entire perceptual system that are linked to specific forms of contextual connections in multiple cortical areas and these processes likely present a basic mechanisms for perceptual consciousness as well as for dissociative phenomena.

In this context, dissociation presents a mental event in which some contextual element is conflicting and although it is a part of the mental scheme at the same time it does not fit into this contextual framework. The processes of contextual interactive competiton suggest that neural entropy and complexity that enables to describe levels of interactions related to order and disorder among neural assemblies involved in cognitive processing in principle could reflect typical attentional changes related to divergent conflicting components in dissociative states (Li and Spiegel 1992; Bob 2003; Bob and Svetlak 2011; Bob et al. 2011). In this sense, the dissociative states are related to mental events that bind free energy into two or more conflicting mental states and behavioral patterns that lead to increased psychological tension and stress.

2.2 Conflicting Information and Dissociated Mind

In agreement with the thermodynamic concept of mind proposed by Freud in his "Project", dissociation or splitting of the mind is typically related to mental divergent components that determine mental and neural disunity. The conflict disables to use the intentional energy which based on thermodynamical principles means that energy that cannot be used for intentional movement and behavior increases entropy that manifests as increased disorder in the system. In this context, typical consequence of a conflict in which an amount of intentional energy cannot be bound to intentional behavior causes dissociation as a defensive response leading to mental disorganization that enables reduction of psychological tension. Dissociation is typically related to long-term or acute stress that significantly disturbs self-concept, identity, memory and perception of the external world. On the level of personal experience dissociation is related to mental fragmentation or splitting due to the conflict and confusion that disturbs meaningful intentional behavior. Consequently these alterations may be linked to great and abrupt changes in patterns of neural activity that may cause dissociation of certain external and internal stimuli out of awareness, and the lack of the self-representation which may lead to a distinct state of dissociated or divided consciousness (Crawford 1994; Rainville et al. 1997, 2002; Villemure and Bushnell 2009; Bob 2007, 2008a; Vermetten and Douglas 2004). These dissociated self-representations have sensory, emotional and cognitive elements that may be misinterpreted and experienced as external reality for example during projection or transference when inner psychic states are interpreted as external parts of other persons or during hallucinations, when certain internally generated voices or images are interpreted as sensory signals from the external world (Feinberg 1978; Feinberg and Guazzelli 1999; Ford et al. 2001; Tsakiris et al. 2005).

Cognitive and affective representations of one's identity or the subject of experience present a basis for self-recognition as a specific cognitive process typically involving conscious experience and interpretation activity. Disruptions of these self-interpretation processes likely represent a neurophysiological substrate for the process of fragmentation of consciousness because of misattribution of certain inner states that may be interpreted as external objects because they are "disowned" and dissociated from consciousness. Psychological or physical stimuli leading to hopelessness and conflicting situation without known solution that do not fit into current cognitive scheme may lead to dissociation and "depersonalization" of certain perceptions, emotions and cognitive strategies that create discrete "ego-states" [or "alter-personalities" in dissociative identity disorder] which are divided from predominant state of consciousness (Bob 2008). All these processes of disrupted awareness and conscious integrity likely are related and represented by similar disruptions on the brain level, and dissociation, in principle, may be explained by various levels of disturbed binding and brain coherence that may negatively affect usual patterns of connectivity, complexity and synchronous activity constituting adaptive integrative functions of consciousness (Bob 2011).

History of research about dissociated mind started Hughlings Jackson (Ellenberger 1970). Jackson studied the so called "dreamy states" frequently related to temporal lobe seizures. His descriptions of the dreamy states first appeared in 1870s and included their typical symptoms such as hallucinations, strangeness, unreality of things such as derealization or depersonalization, double consciousness (looking at one's self through the eyes of others), déjà vu and jamais vu, various autobiographical memories including flashbacks and other symptoms currently described in medial temporal lobe epilepsy (Jackson 1931; Meares 1999; Hogan and Kaiboriboon 2003, 2004).

In his writings Jackson reported numerous cases of the dreamy states and described close association of the dreamy states with immediate onset of the epileptic discharges (Jackson 1931). Recently, similar cases reported Vignal et al. (2007), who studied spontaneous and provoked dreamy states using electrical discharges localized within mesial temporal lobe structures and found that early spread of the discharges to the temporal neocortex prevented the occurrence of the dreamy state. In agreement with further evidence emphasizing the central role of the amygdala and hippocampus (both left and right) in the recall of recent and distant memories they also found, that the pathological activation of the amygdala and hippocampus during seizures may trigger memory recall. Also Penfield found that visual hallucinations were elicited by electrical stimulation in sites widely distributed over the temporal neocortex (Penfield 1967; Penfield and Perot 1963). Penfield reported that these experiential hallucinations replying past experiences reflected a bi-directional activation of temporal neocortex involved in memory storage and also centrencephalic system participating in integration of memories. Further research has shown that the dreamy states may be elicited by stimulation of the medial temporal lobe structures (Ferguson et al. 1969; Halgren et al. 1978; Gloor et al. 1982; Gloor 1990; Bancaud et al. 1994).

Jackson also proposed interpretation of the dreamy states as a release of lower level of brain functions because of inactivated inhibition of highest functional levels caused by the epileptic discharges and not as a direct consequence of epileptic discharges. For theoretical explanation of his clinical data Jackson used the principle of "dissolution" that causes automatic processes with less voluntary control and less complex mental structure than occurs in a normal state mind because there is a loss of later developed higher nervous functions leading to a dysregulation and exaggeration of more primitive functions (Ellenberger 1970; Meares 1999). This interpretation of the dreamy states is in agreement with findings that functional de-activation of the amygdala and hippocampus by epileptiform activity leads to liberation of neocortical structures (Halgren et al. 1978; Bancaud et al. 1994).

Following these Jackson's findings Theódule Ribot in his works on diseases of memory introduced an important concept of retrograde amnesia that was later formulated as "Ribot's Law". This states that brain damage has greater influence on recent than on remote memories, i.e. that more recent memories disappear before the earlier ones. The Ribot's Law opened new era in the study of disorders of memory such as amnesia, multiple personality disorder, hysteria and other dissociative phenomena (Ellenberger 1970). He adopted the principle of evolution and dissolution proposed by Jackson (Ellenberger 1970) and applied this principle to the psychopathology of memory and will, which means that similarly like memory also voluntary activities disappear before the earlier ones. This principle is source for the later formulated theory of psychasthenia by Pierre Janet which preceded the formulation of the theory of dissociation (Ellenberger 1970; Meares 1999).

Ribot's findings developed his follower Pierre Janet, who proposed the concept of dissociation as a specific defensive response leading to memory loss (Ellenberger 1970). Pierre Janet, in his work about psychological automatisms, defined the process of loosening associations using word 'désagrégation' that was later synonymously used with the term "dissociation" and became known mainly through the works of William James and Morton Prince (Ellenberger 1970). According to this Janet's description, dissociation means deficit of the associated system that creates the secondary consciousness (Janet 1890; Ellenberger 1970; van der Hart and Friedman 1989; van der Kolk and van der Hart 1989). Following hypnotic experiments with his teacher Charcot, Janet found that people under hypnosis experienced exceptional states of divided consciousness which in some cases manifested as "different personalities" (Janet 1890; Ellenberger 1970). Several later studies reported similar data as Janet described. For example, Bowers and Brecher (1955) reported interesting material involved in the emergence of multiple personality structure under hypnosis. The authors conclude that this structure was not produced by the hypnosis, but preceded the beginning of the hypnotic work. The patient in the case under discussion had not shown the multiple structure in clinical and psychological examinations prior to the hypnosis. In his conscious state the patient was not aware of his three underlying personalities, each of which reported distinctive dream material and Rorschach responses.

Taken together these data suggest that binding between conscious contents and self-function observed in these cases constitutes the self-representational dimension of consciousness which is characterized by interpretation of certain inner states of own body as mental and somatic identity, while other bodily signals are interpreted as perceptions of the external world. Self-representations that are currently not accessible to the dominant interpreter's access are dissociated and may be defined as subliminal self-representations (Bob 2008).

This line of investigation later appeared also in works of Sigmund Freud and Joseph Breuer who considered secondary consciousness in "Studies in hysteria" (Breuer and Freud 1895). In this study, Breuer and Freud developed concept of splitting as a form of dynamical dissociation that leads to shifts of mind during the time due to a conflict of opposing mental forces (Ellenberger 1970; Breuer and Freud 1895). Based on these principles they conceptualized pathological conditions observed in conversion phenomena as a consequence of repression and according to them "dissociated states" are elicited by the repression of the libido energy. In this context, they also proposed new conceptual framework for understanding of the mind-body problem in which mental and somatic factors are closely connected and understood as different aspects of a unity (Ellenberger 1970; Breuer and Freud 1895; Briquet 1859; Mace 1992; Rofe 2008).

The term repression is historically linked to the term abreaction (van der Hart and Brown 1992) defined as an emotional release or discharge after recalling a traumatic experience that has been repressed because it was consciously intolerable (American Psychiatric Association 1980). Breuer and Freud in their "Studies of Hysteria" proposed the concept of abreaction as an emotional discharge related to release of repressed traumatic memories into conscious awareness (Breuer and Freud 1895; Ellenberger 1970). Later data show that psychotherapeutic work that enables integration of the released traumatic experience into predominant cognitive scheme and contextual framework may have positive effects (Brown and Fromm 1986; Horowitz 1986; Ross 1989; Van Der Hart and Brown 1992; Putnam 1992; Braun 1986).

In the same context, Janet elaborated the concept of dissociated mind in his work Psychological Automatism and other works (Havens 1966; Janet 1886, 1890; van der Hart and Friedman 1989). He dealt with psychological phenomena often observable in hysteria, hypnosis and states of suggestion or possession and found that during complete psychological automatism related to psychological regression consciousness is totally dominated by repeating past experiences, such as in somnambulism or hysterical crises (Janet 1890; Ellenberger 1970; van der Hart and Friedman 1989). Janet also described partial automatism in which only a part of consciousness is dominated and proposed that in these states play an important role unconscious mechanisms related to traumatic experiences that repress conscious control and perception and called them subconscious fixed ideas (Janet 1890; Ellenberger 1970). In the recent literature the fixed idea is defined as a formation of new spheres of consciousness around memories of intensely arousing experiences with a high emotional charge, which organize cognitive, affective and visceral elements of the traumatic experience while simultaneously keeping them out of conscious awareness (van der Hart and Friedman 1989). The fixed ideas may emerge in many forms of psychopathological or somatoform symptoms, which may be understood as a representation of psychological trauma when a fixed idea is transformed into hallucinations and dissociative episodes, for example during paroxysms or hysterical attacks (van der Hart and Friedman 1989). Fixed ideas may be also represented in the form of dreams or during hypnosis as a secondary consciousness. Several later studies reported similar data as Janet described. For example, Salley (1988), who studied his patient Frank with multiple personality and found that his alter personalities emerged on parallel levels on the one hand in dreams and on the other in hypnosis. Frank's multiple organization began at the age of 6 years when he lived with his remaired mother who had an alcoholic man, who abused Frank physically and emotionally. In his anamnesis was a wide spectrum of dissociative symptoms such as a history of blackouts, amnesia for certain experiences, fugues, abrupt personality changes and hysterical conversions. Frank had a long history of frequent seizures with amnesia. The memory typically returned within a few days after the seizure. During hypnotherapy suggestive methods aimed to uncover lost memory were used. During hypnotic trance appeared Frank's subpersonality who identified himself as Self, a protector of Frank. Self in somnambulistic trance, explained that the seizures resulted from a struggle between Frank and Self. Self stated that his only line of communication with Frank was through dreams and that he would create a dream that would explain to Frank the functions of the seizures. Out of trance, Frank as was typical, had no memory of what had occurred in hypnosis. That night Frank dreamt that he was standing on a pedestal and two voices were shouting at him: one voice shouting "Yes!" and the other "No!" The vibrations from the shouting were so intense that the pedestal began to shake and split open, whereupon he fell to the ground shaking. Free association to the elements of the dream led Frank to relate the shaking to his seizures and the screaming to internal conflict. In the 2 years since he had this dream, he has experienced no recurrence of the dissociative seizures which suggest that the dream reflected communication between dissociated and conflicting parts of his mind (Salley 1988). The Salley's case study and other documented dream works with patients suffering from multiple personality disorder and other traumatized patients present important data for research of parallel levels between dream and hypnotic states suggesting that individual alter personalities may create dreams that provide therapeutically meaningful information, enable communication among dissociated parts of the personality and provide access to underlying personality structure (Ferenczi 1934; Levitan 1980; Putnam 1989; Jeans 1976; Marmer 1980a, b; Hartmann 1998; Guralnik et al. 1999; Bob 2004).

A characteristic feature of these mental conflicts and extreme stress is also a lowering of the mental level (abaissement du niveau mental), which is manifested by increased dissociation and mental depression connected to the reduction of psychological tension. Mainly these changes linked to conflicting streams of information specifically influence integrative functions of consciousness and during transient periods related to actual experience of aversive events lead to a greater allocation of attention which leads to discontinuous attentional shifts (Guralnik et al. 2000; Vermetten and Bremner 2004; Bob 2008) in clinical forms of dissociation or during hypnosis (Faymonville et al. 2006; Cojan et al. 2009).

According to modern definition, dissociation presents a special form of consciousness in which events that would ordinarily be connected are divided from one another (Li and Spiegel 1992). According to DSM-III-R and DSM-IV dissociation is defined in as "a disturbance or alteration in the normally integrative functions of identity, memory or consciousness". Other usual definition is based on disintegration of consciousness as an inability to integrate some psychic contents into the consciousness (Bernstein and Putnam 1986). Nemiah (1981) defined basic features of dissociation as alterations of identity related to disturbances of memory of an individual during dissociative states. Changes of identity may appear as depersonalization or in serious cases it may lead to dissociative identity disorder (or multiple personality disorder in older terminology) and typical consequences of the contextual disturbances of mental associations are for example psychogenic amnesia or changes in interpretation of the external world such as derealization or hallucinations (Spiegel and Cardena 1991; Putnam 1989; Bob 2008). Dissociation has also typical somatoform symptoms such as alterations in sensation of pain (analgesia, kinesthetic anesthesia), painful symptoms, perception alterations, motor inhibition or loss of motor control, gastrointestinal symptoms (Nijenhuis et al. 1996, 1998, 2000, 2004) and dissociative seizures (Brown and Trimble 2000; Kuyk et al. 1999). For example, Röder et al. (2007) described that during hypnotic depersonalization they observed reduced activity in sensory and affective pain-related areas, reflecting the diminished intensity of the perceived pain. They also found that networks in cortical and subcortical areas related to the perception of the own body was during depersonalization also less responsive, which might point to a specific neural mechanism underlying the out-of-body experiences.

Although dissociation presents a concept proposed for description of psychopathological phenomena, recent findings show that dissociation is not only pathological but also involves some adaptive functions. Dissociation occurs within a continuum of dissociative symptoms also in the normal population and only abnormal forms of dissociative symptoms are attributed to mental disorder (Putnam 1989). Putnam, in agreement with historical tradition, in his studies of dissociative reaction proposed that the major part of dissociative disorders was induced by traumatic events (Putnam 1989, 1997) and also other recent studies indicate that dissociative disorders are mainly induced due to a traumatic event (Sar and Ross 2006; Brewin 2007; Bob 2008). On the other hand some studies also show that brain insult, injury or other organic brain disease may play a role in this process (Kihlstrom 2005; Spiegel 1997) and in ICD-10 is also defined the organic dissociation, induced by a variety of conditions affecting cerebral function (Good 1993). The concept of organic dissociation is based on evidence that dissociative symptoms and disorders including amnesia, fugue, depersonalization, multiple personality, automatisms, and certain furors can be induced by a variety of medications, abuse of drugs, and medical illnesses or conditions affecting CNS functions and it is important to note that organic dissociation can be distinguished from intoxication, amnestic disorder, and delirium (Good 1993).

According to recent evidence the most important traumas inducing dissociative disorders originate in childhood and are most frequently related to physical or sexual abuse with following development of symptoms often after many years (Teicher et al. 2003, 2006) but also traumatic events after serious accidents or natural disasters may cause dissociative symptoms or disorders (Spiegel and Cardena 1991; Putnam 1997). For example, Chu and Dill (1990) investigated dissociation by means of Dissociative Experiences Scale (DES) in 98 females and found significantly higher dissociation in patients who were exposed to emotional or physical abuse. Coons et al. (1989), in their study of prevalence of traumas in childhood and adult clinical population, found that 100% patients with atypical dissociative disorders and 82% diagnosed as psychogenic amnesia documented physical or verbal abuse or neglect in childhood but also about half of the patients experienced also significant trauma in adulthood.

In this context, there is evidence that exposure to a significant psychological stress preserves or even enhances memory for emotional aspects of an event, and simultaneously disrupts memory for non-emotional aspects of the same event (Payne et al. 2006). For example, Briere and Conte (1989) have documented that 59.6% from the group of 468 patients with proven history of sexual abuse in childhood were not able to remember the episodes of abuse from the past. Similar results, found in the group of 87 children aged 7–15 years exposured to a traumatic event requiring hospitalization, indicate that specifically children who showed temporal disorganization, but not absence of emotion or dissociative amnesia, in narrative themes were more likely to report concurrent subsyndromal PTSD symptoms at 4–7 weeks post-trauma (Kenardy et al. 2007).

Well documented evidence also shows that individuals who are victims of a trauma in many cases are unable to register pain (for example during self-injury) or painful affects (Butler et al. 1996; Frankel 1996; Agargun et al. 1998; Ebrinc 2002; Trief 1996; Saxe et al. 2002; Russr et al. 1993; Orbach et al. 1997) and that patients with dissociative disorders frequently report amnesia for self-injury (Saxe et al. 2002; Putnam 1989). Other interesting case studies describing multiple personality organization have been also reported in chronic pain patients with dissociative identity disorder (Frances and Spiegel 1987; Livengood et al. 1994; McFadden and Woitalla 1993; McFadden 1992; Fishbain et al. 2001; Bob 2008). For example, a 34-year-old woman with chronic pain of the right wrist who developed dissociative identity disorder (Frances and Spiegel 1987) or a male who demonstrated another personality during the period of a severe untreated headache (Livengood et al. 1994).

Conversely, a most common somatic complaint in patients with dissociative identity disorder headaches were reported (Greaves 1980; Bliss 1980; Coons et al. 1988; Coons et al. 1988; Coons 1980; Putnam et al. 1986) and symptoms of headaches in dissociative identity disorder patients have been described as extremely painful during the switching of personalities (Coons et al. 1988; Coons 1980; Larmore et al. 1977; Packard and Brown 1986). For example, several reported cases suggest that patients with dissociative identity disorder can eliminate pain in the primary personality by displacing it into other alters (Watkins and Watkins 1990). Serious symptoms of headaches have also been reported in association with a sudden unexpected travel away with amnesia and confusion about personal identity in patients with dissociative fugue (O'Brien 1985; Fishbain et al. 2001).

In addition there is evidence that not only effects of physical abuse, sexual abuse, or witnessing domestic violence but also parental verbal aggression as a specific form of abuse may cause dissociation. Teicher et al. (2006b) collected data from 554 subjects 18–22 years of age (68 % female) who responded to advertisements. They used Verbal Abuse Questionnaire to assess exposure to parental verbal aggression and measured also dissociation, symptoms of limbic irritability, depression, anxiety, and anger-hostility. They found that exposure to verbal abuse alone and witnessing of domestic violence had moderately strong effects on dissociation ratings and that subjects exposed to both verbal abuse and domestic violence (but no other form of maltreatment) had Dissociative Experience Scale scores 4.5 times as high as those of the nonabused subjects, which implicate that parental verbal aggression is a serious form of child abuse and maltreatment that influence dissociation, symptoms of limbic irritability, depression and other psychiatric symptoms (Teicher et al. 2006).

Close relationship among symptoms of traumatic stress, dissociation, limbic irritability and depression has also been documented in depressive patients (Bob et al. 2010b). The authors found that increased presence of the traumatic stress symptoms in 113 unipolar depressive patients treated with SSRIs is associated with significantly more severe symptoms of dissociation, depression and limbic irritability in comparison with 86 healthy controls. Results indicate that traumatic stress symptoms in depressive patients have significant association with symptoms of dissociation, depression and limbic irritability in the both groups. In this context, there is growing evidence that traumatic experiences present a very important factor in many psychiatric disorders and that dissociative symptomatology often occurs due to child abuse especially in cases of chronic emotional, physical or sexual abuse (Spiegel and Cardena 1991; Putnam 1997; Teicher et al. 2003; Sar and Ross 2006).

These changes in self-perception, self-awareness and self-representation related to dissociative states are likely consequences of profound changes in affect state, memory and sense of identity in response to extreme stress (Saxe et al. 2002), which is specifically related to changes in patterns of neural activity (Crawford 1994; Vermetten and Bremner 2004) that lead to alterations in neural binding (Bob 2011). These profound alterations also lead to significant changes in interpretation of inner signals that have consequences for sensing of own identity and external world (Bob 2008). According to Helen Watkins' concept (Watkins 1993) these dissociated self-representations emerge as ego states [even in non-multiple subjects] and represent organized cognitive structural system of segments of the personality often similar to true multiple personalities. The explanatory model is also based on reported data that these dissociated self-representations were reported in normal people in hypnosis and during dreaming in subjects with multiple personality disorder (Watkins and Watkins 1979–1980; Bowers and Brecher 1955; Gabel 1989; Watkins 1993; Lynn et al. 1994; Merskey 1992; Rickeport 1992; Barret 1995, 1996; Bob 2004). Together these findings indicate that consciousness in dissociative states may significantly change which is mainly based on mechanisms of exogenous and endogenous attention. Modulation of attention in dissociative states is coupled to global changes in subjective experience that is specifically focused on certain internal and external stimuli, mainly due to stressful and traumatic events that significantly influence regulation and monitoring body, mental state, experiencing of the self and underlying process of self-representation (Bob 2008). A level of conscious integration or disintegration may change due to various conditions related to experimental cognitive manipulations, hypnosis or stressful experiences which may lead to dissociation of consciousness and influence accessibility of various mental contents into the consciousness (Baars 2002; Diaz and McCarthy 2007; Melloni et al. 2007).

In this context, it is possible to describe a threshold between consciousness and the unconscious processes that defines potential conscious experiences or secondary consciousness existing behind the threshold (Bob 2003a; Ellenberger 1970; Jung 1972) that may become a part of conscious experience (Stross and Shevrin 1968, 1969; Bob 2003a; Shevrin 2001; Kihlstrom 1987, 2004). The threshold is in principle defined by findings indicating that cerebral cortical activities occurring as a response to a somatosensory stimulus must proceed for about 500 ms in order to elicit the conscious sensation (Libet 1979; Libet et al. 1964, 1967, 1991, 1992). With respect to this finding, several research studies have examined the role of perceptual awareness in processing of emotional pictures that were performed by a short projection of an image (Pessoa 2005; Wiens 2006; Wortman et al. 1992; Karremans et al. 2006). Further studies confirmed that responses to emotional stimuli in the human amygdala occur in the absence of awareness (Pessoa 2005; Wiens 2006; Wong et al. 1994, 2004) and also show that unconscious attentional orienting is specifically mediated by the amygdala, subcortical retinotectal pathway, superior colliculus and several other structures (Mulckhuyse and Theeuwes 2010). Taken together these findings and other reported studies provide evidence of subliminal perception and information processing (Poetzl 1960; Marcel 1983; Reanault et al. 1989; Roediger 1990; Crick and Koch 1995; Brazdil et al. 2001; Bunge et al. 2001; Bob 2003a; Liddell et al. 2004; Gawronski et al. 2006; Kanwisher 2001; Shevrin 2001; Kihlstrom 1987, 2004) and similar findings have been reported also in specific conscious states in split-brain patients where each hemisphere executive control over one side of the body, based on conscious inputs, may be limited to half of the visual field (Gazzaniga and Sperry 1967; Sperry 1968).

Taken together these fluctuating states of consciousness and self-awareness within one human brain likely present a form of personality-state-dependent information processing typically manifested in the brain areas and networks involved in experience of the self that were reported also by some neuroimaging data (Reinders et al. 2003). In this context, Putnam (1997) based on his clinical experience and reported data proposed that observable differences between two mental states related to specific affective state and self-representation produce differential behavioral patterns. These specific patterns of mental and neural activity produce discrete "state dependent" behavioral states of which alterations are manifested as "switch" that represents abrupt change in the values of state defining variables, for example rapid shifts in mood and behavior that could be related to nonlinear dynamic processes (Wolff 1987; Kaplan and Sadock 1991; Putnam 1995, 1997; Gottschalk et al. 1995; Maaranen et al. 2005). Empirical background and definition of the discrete behavioral states originates from the study of infant mental states characterized by basic set of behavioral states that tend to increase their interconnections during development which leads to typical behavioral states in adulthood (Putnam 1997).

Using this model of discrete behavioral states might be in principle possible to consider various levels of order and disorder in the human mind and the brain using some physical concepts of nonlinear dynamics describing levels of entropy and complexity and also to describe some specific relationships between mind and brain using physical and mathematical methods. For such description in principle it is needed to find basic structural components that enable define specific space for their interactions in mathematical language and formulate basic principles of these interactions that enable to understand why certain interacting states may create coherent wholes as a basis for integrative experience or lead to disintegration that determines discontinuous shifts among mental states.

In this context, several findings also suggest that these deficits of conscious integration related to dissociation and increased complexity are conceptually in agreement with theoretical concepts of neural processing known as parallel distributed processing (PDP) or connectionist models of memory (Yates and Nasby 1993; Mc Clelland and Rumelhart 1986). From this point of view dissociation can be described as a kind of divided or parallel neural process where several systems may have some independence (Li and Spiegel 1992; Butler et al. 1996). In this context, dissociation in principle could be understood using the concept of brain complexity describing a level of parallel distribution in neural processing (Butler et al. 1996; Li and Spiegel 1992; Mc Clelland and Rumelhart 1986; Yates and Nasby 1993).

Recent more general view of dissociation and splitting that in principle enables to connect different frameworks proposed by Janet and Freud (Ellenberger 1970) suggested Kohut (1971), who defined horizontal splitting leading to "repression" characterized by a period of dominance of certain attitudes and on the other hand vertical split in which the mental contents and intentions are the side-by-side due to incompatible psychological attitudes.

In this context, the difference between vertical splitting and horizontal splitting is likely defined by a time needed for attentional shift between the two or more conflicting contextual frameworks. The time defines period of dominance for the currently predominant interpretation, which is related to specific contextual and attentional focus. Longer periods are present in the cases of horizontal splitting (or dissociation) defined by Freud as repression, which determines changing and shifts of mental states characterized by opposite attitudes and intentionality during the time, for example in bipolar disorder in the form of manic or depressive mood or in borderline personality disorder where the splitting typically alternates in opposite attitudes (for example discontinuous shifts between love and hate) with respect to the same person or an object. On the other hand short or very short periods of attentional and contextual shifts lead to decreased period of dominance of certain attentional focus, contextual framework or interpretation which determines that opposite attitudes and intentionality are present or experienced as rapid shifts with very short intervals or almost at the same time. This extreme intrapsychic conflict causes that the attitudes and intentionality are directed in opposite ways almost at the same time, which leads to vertical splitting that in an extreme form is present in schizophrenia and manifests as splitting of associations that severely disturbs intentional behavior.

Description of the mental space based on continuous or discontinuous switching from one mental state to another in principle may be understood within the thermodynamic concept of entropy and complexity describing level of independency of discrete behavioral states that produces divergent mental components and increased "neural disunity" as Freud (1875) thought. This level of independency and complexity mainly increases due to mechanisms of vertical splitting (or dissociation) leading to an antagonism among certain interacting behavioral states that cannot create coherent context as a basis for integrative experience which leads to discontinuous shifts among mental states. Consequently these alterations in "self-representation", underlying specific changes in subjective experience, are likely related to specific alterations in "binding" creating neural contextual patterns across various parts of the brain, which implicates that these alterations cause abrupt changes in patterns of neural activity and "neural disintegration".

Chapter 3
Consciousness and Neural Unity

Based on the innate structural preferences of 'association patterns', similarly as in chemical thermodynamics, there is likely a tendency to create dynamic patterns of the mind with lowest possible level of free energy reflecting feelings of 'satisfaction' and balance that likely also define preferred patterns of the brain states. In this context, it is possible to assume that mental integration and unity existing as a mental potentiality related to 'hope' and 'meaning', could represent 'balanced' state of the mind that focuses its free energy to the external world without a conflict, which as Freud suggested in principle might be linked to 'neural unity'.

The event of attentional awareness is coupled to synchronized oscillatory activity in the gamma-band range that is related to gestalt perception and also memory processing directly linked to a stimulus (or endogenously generated process), and accompanied by increased synchrony between sensory and prefrontal regions (Fell et al. 2003; Kaiser and Lutzenberger 2003; Womelsdorf and Fries 2007; Sauseng and Klimesch 2008; Fries 2009).

In this context, there is evidence that consciousness might help to integrate brain functions that are separated and independent (Freeman 1991; Varela et al. 2001; Kanwisher 2001; Baars 2002). These findings also suggest that consciousness might be a gateway to brain integration that may enable access between otherwise separated neuronal functions, for example in cases of executive dysfunctions in patients with dissociative identity disorder, dissociative fugue or during hypnosis (Baars 2002). In these cases a mutual dependence between state of consciousness and executive input has been found and several memories were accessible in some states of consciousness while in others not, which suggest a loss of one executive interpreter's access to conscious events during period of dominance of another interpreter (Hilgard 1986; Putnam 1986, 1997; Baars 2002).

According to recent evidence mental functions are related to specific brain structures, but at the same time there is evidence that mental functions are related to large-scale integration of distributed brain activities that cannot be explained within the concept of brain localization. Historically important concept of the relationship between mind and brain suggested Rene Descartes, who in his works about three

P. Bob, *The Brain and Conscious Unity*, DOI 10.1007/978-1-4939-2700-5_3

centuries ago, looked for "the seat of the soul" within the brain and thought that this special place enables to integrate all mental processing and causality of bodily movements, and the hypothetical center for the unifying local-spatial convergence of neural signals was later called the "Cartesian theater".

A new epoch in this research how the brain reaches its unity, and why it does, started Francis Crick (1994) in his book "The Astonishing Hypothesis: The Scientific Search for the Soul". In this book, Crick proposed mechanisms that the nervous system uses for large-scale integration of distributed neural activities and based on this integration and synthesis as its outcome, creates subjective experience. Although this problem how the brain reaches its integrity and creates subjective experience is still unresolved, there is a predominant opinion that neural integration creating consciousness emerges from a dynamical nucleus of persisting interactions of neural assemblies. Current research findings suggest also other conceptual approaches how to explain brain integration and its unity related to the so called 'binding problem'.

3.1 Brain Specialization Vs. Integration

According to current scientific evidence brain enables to create process called "the neural correlate of consciousness" that transforms neural activity into reportable subjective experiences (van de Grind 2002) and combines the present multimodal sensory information with relevant elements of the spatio-temporal memory from the past experiences. The study of neural correlates of consciousness is traditionally linked to postulate of the brain localization of the mental functions as was proposed Paul Broca in 1860s, who studied brain deficits in aphasic patients and found the brain center for speech production (Broca's area). Historically significant findings on localization of the brain and mental functions as related to electric localization of brain functions reported by Gustav Fritsch and Eduard Hitzig in 1870s (Finger 1994). Other distinguished historical findings on localization of mental functions reported neurosurgeon Wilder Penfield in the 1940s who found that stimulation of the motor cortex of the left or right hemisphere produces effects of muscle contraction on the opposite side. These findings were later developed by Michael Gazzaniga and Roger Sperry in the 1960s in split-brain patients who found brain lateralization of various mental functions and perceptual processes in the left and right hemisphere (Kandel et al. 2000).

Modern findings in this area of research are experimentally based on using neuroimaging techniques particularly using functional magnetic resonance imaging (fMRI) and positron emission tomography (PET) and other methods that enable to measure also subcortical brain structures with high spatial resolution. These findings show that the neural correlate of mental functions even it is possible to measure increase of the brain activity in specific brain areas during various cognitive tasks is not strictly localized in prominent brain areas but it is distributed at the same time in various parts of the brain. Based on these findings consciousness is not strictly localized in some part of the brain and typically is linked to distributed activities that co-exist with localized processes in some distinguished areas of the brain. In this context there is growing evidence that the neural correlate of consciousness likely represents a coherent neural process that connects distributed brain activities into the coherent whole (Crick and Koch 1992; John 2002) which cannot be simply explained only within the paradigm suggesting localization of the mental functions.

In this context major hypothesis is that the neural correlate of consciousness can compare and bind activity patterns only if they arrive simultaneously at the neural correlate of conscious experience (van de Grind 2002) that enables to combine information from each modality distributed into distinct features and locally processed in different relatively specialized brain regions. The information is likely represented and globally integrated by interactions among these regions through various levels of synchronization and coherence within neuronal populations located in multiple brain regions that are linked to large-scale integration or "binding" mechanisms (John 2002; Singer 2001; Crick and Koch 2003; Zeman 2001; Bob 2009). A key question is whether this large scale integration is again more specifically linked to certain more or less localized structure or whether it really presents non-localized phenomenon as proposed Weiskrantz (1998), who suggested that more likely a pattern of activity and large scale integration among regions than a specific particular brain region may present a critical condition for the neural correlate of consciousness. Historical and recent hypotheses dealing with specific brain structures that could have a key role for the existence of the neural correlate of consciousness emphasized a specific influence of pineal gland (Descartes 1975), intralaminar nuclei of the thalamus (Bogen 1995), hippocampus (Gray 1995) and the anterior cingulate cortex (Cotterill 1994).

Recent concepts concerning the integrative role of the anterior cingulate cortex (ACC) seems to support the possibility that there could be some structural localization of integrative functions. Resolving this dilemma could have key consequences for understanding of the neural correlate of consciousness, which although it is spatially distributed and related to large scale integration may have its "extraordinary places" with specific integrative role that enable conscious integration and cognitive functions. Recent findings on ACC specialization suggest a hypothesis that the functional involvement of the ACC in motor control, cognition and motivation states of the organism distinguishes the ACC from other frontocortical regions and that this overlap provides the ACC with the potential to translate intentions into actions (Paus 2001). For example, there is evidence of increased functioning in anterior cingulate cortex in individuals with greater social insight and maturity. Lane and colleagues found that the activity of the anterior cingulate cortex was greater in subjects who had higher levels of social awareness based on objectively scored tests (Lane et al. 1998).

In principle this hypothesis is supported by recent findings of anatomical connectivity of the ACC and its structural particularities, especially the presence of spindle-shaped neurons (Allman et al. 2001; Paus 2001; Posner et al. 2007; Kukleta et al. 2010). The circuitry containing the spindle cells may possess widespread connections with other parts of the brain and also may serve to coordinate the activity of these diverse parts to achieve self-control and the capacity to focus on difficult problems such as emotional self-control and focused problem-solving, error recognition, and various adaptive responses to changing conditions (Allman et al. 2001; Posner et al. 2007). The spindle neurons present a novel evolutionary specialization of the neural circuitry of anterior cingulate cortex (Allman et al. 2001).

Similar conceptual approach was developed by Rueda and Posner (Rueda et al. 2004; Posner et al. 2007), who proposed that major contribution of the ACC to brain functions relies on its ability to regulate information influx through orienting in order to avoid conflicting responses in behavior (Rueda et al. 2004). This specific integrative role of the ACC also seems to be evident from data suggesting that functional interactions manifested on a level of EEG synchrony in the beta-2 frequency band specifically distinguish ACC from other fronto-temporal regions. This preliminary evidence indicates that EEG synchrony in the beta-2 frequency band (25–35 Hz), which is very close to gamma, may distinguish levels of synchrony between ACC and various fronto-temporal areas from levels of synchrony between frontal-temporal regions (Kukleta et al. 2010). Crucial interpretation of this finding is that higher level synchrony of the pairs connecting ACC electrode with a frontal-temporal electrode in comparison to pairs of frontal-temporal electrodes suggest a particular role of the ACC in large-scale communication, which could reflect its unique integrative functions in cognitive processing.

Taken together these data strongly suggest that complex cognitive functions linked to neural correlate of consciousness are organized at a global level that enables to integrate primitive functions organized in localized brain regions (Bresler 2001). In this context, current predominant hypothesis relies on the assumption of the global mode of functioning that is based on large-scale information processing which requires a mechanism of functional integration of multiple disparate neural assemblies and communication in the brain through synchronization related to binding of spatially distributed information (Varela et al. 2001; Fries 2001; Jensen et al. 2007).

3.2 Cartesian Model of the Mind and the Binding Problem

Some ideas about localization of consciousness in the brain are at the focus of scientific thought from the beginnigs of modern science. More than three centuries ago Rene Descartes described the problem of brain localization of consciousness and thought about the pineal gland as "the seat of the soul" (Descartes 1649/1975, p. 31). This specific center of the nervous system that integrates brain activity with subjective experiences was later called the "Cartesian theatre" (Dennett 1991; Crick and Koch 1992). Descartes thought that when we sense only one image with two eyes, only one sound with two ears or only one object by two hands, the sensations from two sources must be fused somewhere (Descartes 1649/1975; Barrera-Mera and Barrera-Calva 1998; Smith 1998). Descartes intuitively postulated that this information is fused and governed by a mechanism in the pineal gland. He believed that the pineal gland is involved in sensation, imagination, memory and the causation of bodily movements, and described the mind as an extracorporeal entity that is expressed through the pineal gland, a small endocrine gland that according to current findings secrets melatonin and is involved in circadian rhythms (Barrera-Mera and Barrera-Calva 1998; Smith 1998).

Descartes used interesting arguments why he thought that the pineal gland has this exceptional role and his reason was that the other parts of the brain are double and related to lateralized structures as we have two eyes, two hands, two ears and all the organs of our outside senses are double even we have unified scene of one particular thing at one and the same moment (Descartes 1649/1975). This simple argument of unified symmetry he used to understand for the integration of duality in unity and used the "clock" metaphor as a basic mechanism of the universe and an explanation for basic functional mechanism of the brain and other physiological functions suggesting that the information integrated in the soul must reflect time synchronization. He thought that when we sense only one image with two eyes, only one sound with two ears or only one object by two hands, the sensations from two sources must be fused due to the time synchronization that represents basic scheme for a particular moment of time. On the other hand recognizing one particular object is not only linked to the object right now but also includes its time existence from moment to moment. This implicates that the information integration is also linked to the course of time that enables recognition and Descartes intuitively proposed that this information is fused and governed by clock mechanisms in the pineal gland.

In his thoughts, Descartes intuitively anticipated the so-called "binding problem" of consciousness which means the question where and how the information from various sensory modalities is integrated into the whole and how this information is specifically linked to previous experience creating spatio-temporal memory. Compared to current findings about the pineal gland, Descartes' intuitive proposal is quite close to recent evidence about molecular mechanisms of brain integration and melatonin. Melatonin plays an important role among endocrine output signals related to circadian information processing as an endogenous synchronizer which is able to stabilize and reinforce circadian rhythms and to maintain their mutual phase-relationship. This integrative process occurs at different levels of the circadian network via gene expression in some brain regions and peripheral structures that enables integration of circadian, hormonal, and metabolic information and creates temporal order of bodily and mental experience (Rutter 2002; Pevet 2006; Saper 2005).

This specific temporal order is reflected in associative process that is necessary for cognition, behavior and all processes of memory consolidation that must preserve all the information in the temporal causal order and synchrony or sequentiality of the internal cognitive maps. In the processes of temporal integration and binding related to consciousness, cognitive processes and memory functions, a specific and major role play neurons of suprachiasmatic nuclei (SCN) of the hypothalamus. The temporal patterns of rhythmicity in the SCN are generated by a gene expression of individual SCN neurons. These individual neural oscillators are organized into a coherent activity of biological clock and facilitate temporal synchronization that produces differentially timed waves specifically targeting the pineal gland and other structures, and control neuroendocrine rhythms (Kalsbeek et al. 2006; Indic et al. 2007; Hamada et al. 2004).

Also other arguments and modern findings show that the pineal gland really has an exceptional role. For example, recent findings suggest that melatonin could be a potential regulator in the processes that contribute to memory formation, long-term potentiation (LTP) and synaptic plasticity in the hippocampus and other brain regions (Lynch 2004; Baydas et al. 2005; Larson et al. 2006; Chaudhury et al. 2005; Ozcan et al. 2006; Gorfine and Zisapel 2007). Melatonin may also inhibit LTP induction through a regulation of signaling pathways downstream of the membrane and NMDA receptor activation and outside of the hippocampus, melatonin may influence rhythms in gene expression and second messenger systems (von Gall et al. 2002; Gerdin et al. 2004; Wang et al. 2005). Electrophysiological studies also have reported that melatonin may regulate the electrical activity of hippocampal neurons (Zeise and Semm 1985; Musshoff et al. 2002) and alter synaptic transmission between hippocampal neurons (Wan et al. 1999; Hogan et al. 2001; El-Sherif et al. 2002). These findings indicate that melatonin can regulate learning and memory through its influence to synaptic connections within the hippocampus undergoing activity-dependent changes in synaptic strength including enhancements in the strength of excitatory synaptic transmission that regulates LTP.

In addition, basic mechanism of melatonin action is that it may interact with both excitatory and inhibitory neurotransmitter systems (Larson et al. 2006; Saenz et al. 2004; Skaper et al. 1998). A mechanism likely underlying the effects of melatonin on synaptic plasticity is a modulation of the intrinsic excitability of hippocampal neurons. Hyperpolarizaton induced by melatonin could reduce long term potentiation (LTP) by inhibiting NMDA receptor activation during high frequency stimulation (Wang et al. 2005). Melatonin application decreases membrane excitability in other regions of the nervous system in part via an enhancement of potassium currents (Wang et al. 2005). Melatonin also may decrease spontaneous action potential generation in the SCN (Shibata et al. 1989; Stehle et al. 1989; Mason and Rusak 1990) through an increase in a potassium conductance and a decrease in a hyperpolarization-activated current (Jiang et al. 1995; van den Top et al. 2001).

Together the current findings indicate that the role of pineal gland is really exceptional. Descartes' arguments that other parts of the brain are double and lateralized and all the organs of our outside senses are double even we have unified scene (Descartes 1649/1975) could in principle be valid. This argument of "unified symmetry" of the brain that he used to understand the integration of duality in unity could be meaningful because when something is divided into duality for its integration it needs something third which is hierarchically different. In principle it could be possible that the brain has integrative "extraordinary places" that are involved in the brain integration more than others.

3.3 Binding Problem and the Brain

As already Descartes proposed the binding of information from various sensory modalities that enables consciousness leads to the unresolved question where and how the information is integrated into the whole and how this information is specifically linked to previous experience in the spatio-temporal memory. An important question for further research of the integration of distributed brain activities is whether the essential integrative role can be attributed to a specific structure in the brain or whether this ability is inherent to the cognitive network as a whole. In addition it is not known how spatial convergency is provided for the synthesis of processed information that emerges and for example, there are only a few neural connections between specific visual areas that correlate with color and motion (Bartels and Zeki 2006; LaRock 2006; Zeki 1994, 2003). For example, visual consciousness requires activity in many areas of the brain and the components of the high-level visual representations, closely linked to focused attention, need to access these structures, form a perceptual object and bind together various features of an object (Zeki and Bartels 1999; Cavanagh 2011).

A key evidence for the binding problem reported studies of primate extrastriate visual cortex which have shown that different neurons within the visual system participate on processing of different features of a seen object (Desimone et al. 1985; Schein and Desimone 1990; Ghose and Ts'o 1997). For example, Desimone et al. (1985) have found that neurons in visual area (V4) and inferior temporal cortex (IT) are sensitive to many kinds of information relevant to object recognition. They also found that special contribution of V4 neurons to visual processing may lie in specific spatial and spectral interactions and that many different stimulus qualities are processed in parallel mode of processing. Similar results reported Ghose and Ts'o (1997), who also found that V4 contains modular assemblies of cells related to particular aspects of processed object-based representations. There is also evidence in research of moving objects that other neurons in the middle temporal area and the medial superior temporal area encode various aspects related to the motion of the stimulus (Treue and Andersen 1996; O'Keefe and Movshon 1998).

Significant contribution to this discussion about mechanisms of large scale integration reported Crick and Koch (1992, 2003) in their studies of the visual consciousness. They proposed that the problem of binding may not be resolved only as a simple consequence of synchronization among large groups of neurons. As a basis for that opinion they emphasize the binding problem of distributed information representing a seen object by groups of synchronized excited neurons that are located at different parts of the brain. This problem has emerged in connection to findings that features of an object such as color, shape, texture, size, brightness, etc. produce activity in separate areas of the visual cortex (Crick and Koch 1992; Felleman and Van Essen 1991; Singer 1993, 2001). In context of these findings the hypothetical center for brain information convergency that enables perceptual consciousness and conscious experience was termed "Cartesian theatre" (Crick and Koch 1992; Dennett 1991). Recent neuroscience, however, has not located a distinct place in which distributed information in the brain comes together. Additionally, there is evidence that neocortical processing is distributed during all sensory and motor functions (Singer 1993, 2001).

The predominant view in neuroscience of consciousness is that neuronal synchronization is a phenomenon that is necessary for the large scale integration of distributed neuronal activities. There is increasing experimental evidence that coherent neuronal assemblies in the brain are functionally linked by phase synchronization among simultaneously recorded EEG signals and that this time-dependent synchrony between various discrete neuronal assemblies represents neural substrate for mental representations such as perception, cognitive functions and memory (Varela et al. 2001; Lachaux et al. 1999). These functions are related to distributed macroscopic patterns of neuronal activity which involve multiple neuronal subsystems bound into a coherent whole (Braitenberg 1978; van Putten and Stam 2001). According to recent data, a mechanism that enables binding of distributed macroscopic patterns of neuronal activity, represented by neural assemblies, into the coherent whole is still unresolved and represents a fundamental problem in neuroscience [i.e., the binding problem: how the brain codes and integrates distributed neural activities during processes connected to perception, cognition and memory] (Woolf and Hameroff 2001; Lee et al. 2003; Arp 2005; Fidelman 2005; Velik 2010). The theory of feature binding originates with distributed coding and states that neurons involved in the processing of a single object will tend to synchronize their firing, while simultaneously desynchronizing their firing from the remaining neurons not involved in the processing of the object (von der Malsburg and Schneider 1986). An essential feature of neuronal assembly coding is that individual neurons or subsystems can participate at different times in an almost unlimited number of different assemblies (Sannita 2000; Varela et al. 2001). The same neurons can participate in different perceptual events and different combinations of these neurons can represent different perceptual objects. Synchronization of these different perceptual objects is related to the integration of perceptions into a coherent whole (Singer and Gray 1995). As a candidate mechanism for the integration or binding of distributed brain activities is the so-called gamma activity—high frequency oscillations of 40 Hz, but often varying from 30 to 90 Hz. This activity occurs synchronously across brain regions and underlies the integration of diverse brain activities (Singer and Gray 1995).

In this context, majority of recent studies on neural correlate of consciousness have focused on EEG analysis and observed functionally relevant periods of synchronization mainly in gamma frequency band in various species and brain structures during attention, perception, motor and memory tasks (Singer 2001; Lee et al. 2003; Jensen et al. 2007). Together these findings suggest that a candidate mechanism for the integration or binding of distributed brain activities is the gamma activity (mainly about 40 Hz). The suggested mechanism of the gamma waves is that the wave that originates in the thalamus repeatedly oscillates 40 times per second in the brain back and forth, which enable to different neuronal circuits to enter into synchrony with the perceptual information that is processed in thalamus. This integrative process enables to these simultaneously active neuronal clusters oscillate together during transient periods of synchronized firing and this coherent whole enables to connect various memories and associations involved in the process and generate a coherent process of perception, cognitive processing, memory and consciousness (Buzsaki 2006).

Through this mechanism gamma activity occurs synchronously across brain regions and underlies the integration of diverse brain activities (Singer and Gray 1995). Crucial result of these studies is a direct link between visual perception and gamma synchrony in the cat visual cortex reported by Eckhorn et al. (1988). Following this finding, functional significance of synchronous gamma activity in selective attention, perceptual processing and recognition was repeatedly demonstrated in animal and human studies (Fries et al. 2001; Meador et al. 2005; Rodrigez et al. 2004; Jensen et al. 2007). Together these data strongly suggest that complex cognitive functions are organized at a global level that enables to integrate primitive functions organized in localized brain regions (Bresler 2001). In this context, current predominant hypothesis relies on the assumption of the global mode of functioning that is based on large-scale information processing that requires mechanisms of functional integration of multiple disparate neural assemblies (Varela et al. 2001; Fries 2001; Jensen et al. 2007).

Although the majority of research on feature binding has focused on synchronous gamma activity, there is evidence that synchronous activities in other frequency bands may also participate in functional integration of distributed neural activities into the coherent whole (Bressler et al. 1993; Lee et al. 2003).

Following these findings Dennett (1991) proposed "a multiple drafts" theory of consciousness model that does not define consciousness as a unitary process but rather a distributed one. Instead of a single central place i.e. "Cartesian theatre," there are various events of content-fixation that occur in various places at various times in the brain (Dennett 1991, p. 365). The evidence for this view of consciousness presents a whole series of experimental results in cognitive neuroscience and psychology (LaRock 2006; Varela and Thompson 2003; von der Marlsburg 1996, 1999; van der Velde and de Kamps 2006; Zeki 2003).

Chapter 4
Stress, Conflict and the Brain

Even we do not know how the nervous system reaches its unity, integrates distributed neural activities and creates subjective experience, there is evidence that disruptions in functional connectivity and distribution of information are related to perceptual and cognitive states producing disintegration of the mind and 'intrapsychic' conflict. These conflicting activation patterns likely create stress-related subjective experiences and produce neural interference, temporal disorganization and disturbed functional integration of distributed neural activities, which in principle might be explained as disturbed 'neural unity' as Freud suggested in his "Project for a scientific psychology".

Current data provide also evidence that extreme cases of mental dissociation or splitting produce increased disorganization and brain entropy related to disturbed information processing (and possibly a loss of information that in principle increases a system entropy and disorganization). A main aspect of neural disorganization is disturbed large scale communication in the brain that represents a form of "neural disunity" related to cognitive conflict and repressed intentionality. In the light of current findings, Freud's "Project" focused on thermodynamics of the mind in fact provides intuitive view that the thermodynamic approach is closely linked to intentionality and information, and that stress and conflict related to this "intentional" energy causes increased entropy in the system.

Although monitoring and experiencing stress presents a global brain phenomenon, a very important role in the processing of stressful and conflicting experiences likely plays anterior cingulate cortex (ACC) which is significantly involved in cognitive functions related to processing and monitoring of conflicting information that is typically related to higher ACC activation (Bunge et al. 2001; Paus 2001; Bob 2008). Most likely major contribution of the ACC to brain functions relies on its ability to regulate information influx through orienting in order to avoid conflicting responses in behavior (Rueda et al. 2004). In addition, specific integrative role of the ACC also suggests a conceptual proposal of the central autonomic network that includes structures connecting brain structures with autonomic nervous system. This network mainly include ACC, insula and medial temporal lobe structures

P. Bob, *The Brain and Conscious Unity*, DOI 10.1007/978-1-4939-2700-5_4

such as the amygdala and hippocampus that integrate emotional and cognitive information and exert a modulatory role on lower brain centers that control autonomic nervous system and modulates autonomic, neuroendocrine and immune responses (Benarroch 1993; Bob et al. 2009a).

4.1 Neuroendocrine and Immune Response to Stress

A basic feature of stress activation presents corticotropin-releasing hormone (CRH) that regulates peripheral activities of the Hypothalamic-Pituitary-Adrenal (HPA) axis leading to glucorticoid secretion and CRH also influences sympathetic nervous activity and immune responses (Chrousos et al. 1995; Elenkov et al. 2000; Elenkov and Chrousos 2002; Bob et al. 2010c). CRH is released from the paraventricular nucleus of the hypothalamus and influences anterior pituitary gland, which responds by release of adrenocorticotropin hormone (ACTH). The ACTH signal then goes to adrenal glands, which synthesize and release glucocorticoid hormone cortisol that plays an important role in various brain processes related to memory and emotion, and influences also metabolism on various levels where it regulates glucose storage and its utilization (Sapolsky et al. 2000; Miller et al. 2007).

HPA axis is mainly regulated by a negative feedback in which increased levels of cortisol suppress output of CRH and ACTH by acting on glucocorticoid receptors in the hippocampus, hypothalamus, and pituitary gland (Miller et al. 2007). This chronically increased cortisol levels are frequently related to functional deficits in the hippocampus, which lead to decreasing inhibitory control of the hippocampus on the HPA axis and cause a positive feedforward cascade of glucocorticoide levels (Bao et al. 2008).

Recent data indicate that most serious disturbances of HPA axis are caused by traumatic events such as childhood abuse or neglect in the first years of life and often have long-term impact on emotional, behavioral, cognitive, social and physiological functions (Horowitz et al. 1979; Ito et al. 1998; Heim et al. 2000; Orr and Roth 2000; Teicher et al. 2003; Read et al. 2001). Repeated stressors and mainly reexperiencing of the traumatic event in childhood frequently may also cause the delayed effects of severe psychological trauma that lead to enhancement of the self-preservative catecholamine states related to anger, fear, meaninglessness and a blunting of the emotional responses of the attachment behavior associated with dysfunction of the locus coeruleus, amygdala and hippocampal systems (Henry 1992, 1997).

The CRH also directly stimulate cytokines (proteins, peptides, or glycoproteins) representing cell signaling molecules, for example interleukins IL-6, IL-1, IL-12, substance P and also TNF-alpha (Lotz et al. 1988; Leu and Singh 1992; Elenkov and Chrousos 2002). These molecules, predominantly IL-6, IL-1 and TNF-alpha, play an important role in stress activation of the immune system, coordinate the brain-endocrine-immune response and specifically stimulate the hypothalamus, amygdala, and pituitary gland, which in turn stimulate CRH secretion and activate both the HPA axis and the sympathetic nervous system (Besedovsky et al. 1986; Kovacs and Elenkov 1995; Elenkov and Chrousos 2002).

Typical stress response on stress activation related to stimulation of pituitary gland is also linked to melatonin alterations that were found in animals as well as in humans. These studies suggest that the pineal gland may be significantly affected by stress that is consistent with findings that pineal gland expresses high density of the glucocorticoid receptor (Warembourg 1975; Sarrieau et al. 1988; Meyer et al. 1998).

In addition, melatonin receptors are also present in regions that participate in the stress response, such as the hippocampus or the adrenal gland (Musshoff et al. 2002; Torres-Farfan et al. 2003). For example, repeated maternal separation and deprivation caused low blood melatonin levels and a significant negative correlation between blood melatonin levels and spatial memory performance in both male and female adolescent rats that suggest an association between melatonin production and neurodevelopment (Uysal et al. 2005). Further studies also found the interaction between stress and pineal gland (Vollrath and Welker 1998; Simonneaux and Ribelayga 2003) and some electron microscopy studies have found that immobilization stress induces pinealocyte degeneration (Milin et al. 1996). Psychosocial stress also may induce a robust increase of melatonin metabolite 6-sulfatoxymelatonin in subordinate animals (Fuchs and Schumacher 1990).

In humans, stress may cause sleep disturbances, such as insomnia, and reduced nocturnal peak of pineal melatonin secretion that is often present in depressed patients (Jindal and Thase 2004; Brown et al. 1985; Frazer et al. 1986; Pacchierotti et al. 2001). Together these findings suggest that melatonin likely is significantly associated with the regulation of memory, cognition and emotional processes (Laudon et al. 1989; Boatright et al. 1994; Hemby et al. 2003). These findings emphasizing a specific role of melatonin in mechanisms of cognition, memory and stress are also consistent with reported studies that indicate melatonin alterations in psychopathology mainly in patients with depression, schizophrenia, anxiety disorders, eating disorders and also in other mental disorders (Pacchierotti et al. 2001; Bob and Fedor-Freybergh 2008).

4.2 Stress and Brain Derived Neurothrophic Factor (BDNF)

Other basic features of the stress activation are influences of proinflammatory cytokines, such as IL-6, IL-1 and TNF-alpha on deficits in neurogenesis related to changes in neurotrophic factors expression, for example Brain Derived Neurothrophic Factor (BDNF) expression that may explain cognitive and memory deficits related to stress (Schulte-Herbruggen et al. 2005; Maes et al. 2009). BDNF is a polypeptide molecule that influences differentiation and survival of neurons, regulation of synaptic plasticity and connectivity (Bath and Lee 2006; Bramham and Messaoudi 2005) and also growth and differentiation of new neurons and synapses

(Acheson et al. 1995). Its main areas of activity are hippocampus, cortex, and basal forebrain (Yamada and Nabeshima 2003). BDNF also influences excitatory transmission and synaptic plasticity with predominant effective localization of BDNF and its receptor tyrosine kinase TrkB (tropomyosin receptor kinase B) on glutamate synapses (Soule et al. 2006; Bramham and Messaoudi 2005). As a part of these signaling pathways a basic mechanism of BDNF is that it may modulate both excitatory and inhibitory neurotransmitter systems (Savitz et al. 2006). According to several studies BDNF also influences functions of serotonergic and dopaminergic systems (Savitz et al. 2006; Narita et al. 2003; Mossner et al. 2000).

In general the relationship between BDNF and cognition is significantly mediated by influence of stress (Savitz et al. 2006) and especially chronic stress related to excessive release of glucocorticoids from the adrenal gland may cause cell death or atrophy of vulnerable neurons through the cortisol action and inhibitory influence on BDNF synthesis in the hippocampus (Savitz et al. 2006). Together recent findings indicate that BDNF play an important role in stress response and related modification of synaptic plasticity, transmission and memory formation especially in the hippocampus and neocortex, with specific implications for etiopathogenesis of depression, schizophrenia, epilepsy, neurodegenerative disorders and pain sensitization (Binder and Scharfman 2004; Thomas and Davis 2005).

Additionally, there is evidence that decreased BDNF expression may significantly influence reparative processes and neurogenesis of hippocampal neurons (Bremner 2006; Bremner et al. 2008) with resulting structural abnormalities of the hippocampus and other brain regions (Bremner et al. 1995, 2007; Bremner 1999; De Bellis et al. 1999, 2001; Jelicic and Merckecbach 2004; Winter and Irle 2004; Teicher et al. 2006). Meta-analyses of these reported studies show that in PTSD patients hippocampal volumes are significantly smaller compared to controls with and without trauma exposure but trauma-exposed patients without PTSD also have shown significantly smaller bilateral hippocampal volumes in comparison to non-exposed controls (Kitayama et al. 2005; Karl et al. 2006). Direct relationship of structural brain abnormalities with dissociative symptoms and symptoms of traumatic stress suggest data reported by Choi et al. (2009). They studied 16 unmedicated subjects (4 male/12 female subjects, mean age 21.9 +/− 2.4 years) with history of high-level exposure to parental verbal abuse but no other form of maltreatment and 16 healthy control subjects (5 male/11 female subjects, 21.0 +/− 1.6 years). They found that ratings of dissociation, limbic irritability and depression were inversely associated with integrity of the white matter tract in the left hippocampus [measured by fractional anisotropy] and ratings of somatization and anxiety were inversely correlated with integrity of the white matter in the left fornix. Parental verbal abuse score was negatively significantly correlated with white matter integrity in the left superior temporal gyrus (Spearman $r = -0.701$), in the left hippocampus (Spearman $r = -0.801$) and in the left fornix (Spearman $r = -0.524$). Together these findings indicate that stress through molecular level affects structure and connectivity of the brain (Nadel and Jacobs 1998; Payne et al. 2006; Bremner 2006; Bremner et al. 2008).

In general these findings are in agreement with increasing evidence indicating that child abuse and other traumatic stress experiences caused by unescapable adverse physical, emotional or social events represent significant conditions in pathophysiology of various psychiatric disorders (Teicher et al. 2003; Duman and Montegia 2006; Dranovsky and Hen 2006; Bogdan and Pizzagali 2006; Cole et al. 2006; Harkness et al. 2006; Monroe et al. 2007). Early stress may determine developmental abnormalities in the amygdala, hippocampus, cerebellum, anterior cingulate cortex (ACC), corpus callosum and other brain structures that play a critical role in mediating response to stress (Teicher et al. 2003, 2006; Riklan et al. 1977; Putnam 1995, 1997; Bremner 2006). In addition, stress also significantly influences decrease in brain-derived neurotrophic factor (BDNF) in the hippocampus and some other structures that may influence depression and neurodegenerative process (Duman and Montegia 2006; Calabrese et al. 2009; Roth and Sweatt 2011).

4.3 Stress, Gamma Aminobutyric Acid (GABA) and Sensitization

In general, GABAergic neurons provide both inhibitory and disinhibitory modulation of cortical and hippocampal circuits, contribute to the generation of oscillatory rhythms and participate in discriminative information processing such as gating of sensory information, and attentional filtering within the corticolimbic system that is typically affected in several mental disorders for example in schizophrenia (Glenthoj and Hemmingsen 1997; Benes and Berretta 2001; Costa et al. 2004; Gonzalez-Burgos and Lewis 2008; Jacob et al. 2008). Furthermore influence on disturbances in GABA system may also exert increased flow of excitatory activity from the basolateral nucleus of the amygdala (Benes and Berretta 2001). In agreement with this role of GABAergic neurons in cognitive functions several findings also suggest that disturbances in GABA system might be related to stressful conditions and alterations in the dopamine system (Benes and Berretta 2001; Teicher et al. 2003, 2006; Yuii et al. 2007). Taken together, current findings suggest that there is a link between disturbances in GABA system and increased dopaminergic activity that may determine relationships between stress and abnormal neural excitability that specifically influences contents of consciousness and subjective experience and significantly affect brain functions (Teicher et al. 2004, 2006).

According to recent findings increased excitatory neural activity related to stress may also influence a regionally-specific compensatory upregulation of GABA-A receptors in response to decreased GABAergic input in hippocampal pyramidal cells (Heckers and Konradi 2002; Möhler 2006). These stress influences likely may cause changes in GABA postsynaptic receptors that may lead to overstimulation of neurons mainly in the limbic system, resulting in limbic system irritability manifesting as markedly increased prevalence of symptoms suggestive of temporal lobe epilepsy (Teicher et al. 2003, 2006; Post et al. 1995). Repeated stressful events also may determine sensitization that leads to an increase in responsiveness to a stress stimuli resulting from repeated stressors with significantly increased vulnerability to stressors that have more lasting consequences with kindling-like progression (Post et al. 1995; Post and Weiss 1998; Kraus 2000). The kindling-model of stress-related sensitization (Post et al. 1995) seems to be in agreement with suggestive evidence that stress may influence significantly increased occurrence of EEG abnormalities that have been reported in significantly traumatized patients mainly in the frontotemporal region, which consisted of spikes, sharp waves, or paroxysmal slowing, predominantly in the left hemisphere (Teicher et al. 1993, 2003, 2006; Putnam 1997; Ito et al. 1993).

Recent data strongly suggest that early stress may determine limbic irritability and temporal-limbic seizure-like activity (Teicher et al. 2003, 2006; Spigelman et al. 2002) and close link between limbic irritability and cerebellar vermis has been reported (Teicher et al. 2003, 2006; Anderson et al. 2002). Anderson et al. (2002) found reciprocal relationship between activity in the cerebellar vermis assessed by fMRI T2 relaxometry and symptoms of limbic irritability measured by questionnaire LSCL-33 (Limbic System Checklist). They reported a significant correlation between T2 relaxation time and the degree of limbic irritability (LSCL-33) in healthy young adult controls ($r = -0.67$) and also in young adults who had a history of repeated sexual abuse ($r = -0.71$). Teicher et al. (1993, 2003, 2006) also found that adult outpatients with a self-reported history of physical or sexual abuse had increased LSCL-33 scores that were dramatically elevated in patients with a history of combined abuse, both physical and sexual. The results are consistent with findings that cerebellar vermis controls limbic activation and inhibition and also influences the onset and spread of seizures (Heath 1976; Riklan et al. 1976; Schmahmann and Sherman 1998; Teicher et al. 2003, 2006; Schutter and van Honk 2006). These findings suggest that cognitive and emotional dysregulation related to traumatic stress likely is linked to defective inhibitory functions that may also lead to temporo-limbic seizure-like activity. This epileptic-like process may emerge in the form of symptoms similar to ictal temporal lobe epilepsy such as somatic, sensory, behavioral and memory symptoms that may occur also in nonepileptic conditions (Teicher et al. 2003, 2006; Silberman et al. 1985; Roberts et al. 1992; Hines et al. 1993).

Taken together, current findings suggest that there is a link between influence of traumatic stress and psychosensory symptoms related to limbic irritability which are similar to symptoms of temporal epilepsy and likely reflect abnormal neural excitability and disturbances in brain integrative processing that may specifically influence neural correlate of consciousness and subjective experience. In this context, frequent occurrence of epileptic activity and epileptiform abnormalities in dissociative states and disorders suggests that crucial importance for understanding of dissociative processes may have processes related to sensitization and kindling (Goddard et al. 1969) that can potentially explain how epileptic-like phenomena might arise from repeated trauma (Post et al. 1995; Putnam 1997; Teicher et al. 2003). The kindling related to traumatic stress similarly as experimental kindling likely is caused by progressively increasing response of groups of neurons due to repetitive subthreshold stimulation that may later lead to epileptic activity. Repeated stressful events likely may determine sensitization leading to an increase in responsiveness to stress stimuli resulting to significantly increased vulnerability to stressors that have more lasting consequences with kindling-like progression (Post et al. 1995; Post and Weiss 1998; Kraus 2000). In addition increased vulnerability related to sensitization and kindling may cause that the brain becomes also more sensitized for example for depressive states and other symptoms and because of greater sensitivity the onset of future episodes is less related to stressful life events than at the beginning of the disease (Keller 2003; Monroe and Harkness 2005).

The kindling-model of stress-related sensitization (Post et al. 1995) seems to be in agreement with suggestive evidence that stress may influence significantly increased occurrence of EEG abnormalities that have been reported in significantly traumatized patients (Teicher et al. 1993, 2003, 2006; Putnam 1997; Ito et al. 1993). Kindling mechanism caused by stress may involve typical inhibitory failure related to overloading of defensive mechanisms such as denial or "repression" that has been conceptualized for understanding dissociative states (Yates and Nasby 1993). This process therefore leads to similar lack of inhibiton as epilepsy and therefore it may also cause similar pathological electrophysiological changes as found in epilepsy.

Although dissociative disorders cannot be generally explained on the basis of neurological dysfunction, contemporary data provide evidence that temporal lobe seizure activity can produce dissociative syndrome, which is similar to that observed in functional cases (Spiegel 1991). The evidence is based on relatively frequent occurrence of epileptic discharges or epileptiform abnormalities documented in dissociated patients who are victims of child abuse (Coons et al. 1989; Teicher et al. 1993, 2003; Ito et al. 1993; Bob 2003). Typical EEG abnormalities found in these traumatized and dissociated patients often involve temporal or frontal slow wave activity and also may involve frontotemporal spikes or sharp waves predominantly in the left hemisphere (Ito et al. 1993; Putnam 1997; Teicher et al. 1993, 2003). EEG abnormalities also may occur during dissociative states such as depersonalization (Sierra and Berrios 1998), dissociative disorders not otherwise specified and dissociative seizures (the so-called pseudoepilepsy) (Bowman and Coons 2000), religious experiences (Saver and Rabin 1997), out of body experiences (Blanke et al. 2002) and multiple personality disorder (Schenk and Bear 1981; Mesulam 1981; Coons et al. 1982; Benson et al. 1986; Spiegel 1991; Perrine 1991; Hersch et al. 2002). For example, Ahern et al. (1993) examined the relationship of "multiple personality disorder" in two patients with temporolimbic epilepsy to certain types of hemispheric interaction. Both patients exhibited different "personalities" in a characteristic temporal relationship to their seizures. These two patients with temporolimbic epilepsy were considered to be surgical candidates referred for the intracarotid amobarbital sodium procedure. Both patients have demonstrated outbursts of emotional behaviour during inactivation of the left hemisphere. These "different personalities" were known to the patient's families to manifest themselves in the postictal period. These observations suggest that the association of multiple personality and temporolimbic epilepsy is not dependent only on seizure discharges, but may be related also to certain types of hemispheric interaction (Ahern et al. 1993). On the other hand out of body experiences have been observed as a consequence of electrical stimulation (Blanke et al. 2002; Tong 2003). For example Blanke et al. (2002) described the repeated induction of out of body experience by focal electrical stimulation of the right angular gyrus in a patient who was undergoing evaluation for epilepsy treatment. Stimulation at this site also elicited illusory transformations of the patient's arm and legs and whole-body displacements indicating that out-of-body experiences reflect changes in integration complex somatosensory and vestibular information.

Taken together, current findings suggest that there is a link between influence of traumatic stress and limbic irritability which is similar to temporal epilepsy. All these pathological changes specifically influence epileptic or epileptiform synchronization by affecting neurons in some specific structures of the brain and also brain network as a whole. Because of these pathogenic influences some parts of the brain may be damaged which consequently changes neural excitability and also may typically affect processed information and brain integrative processing that may specifically influence neural correlate of consciousness, contextual attentional processing and subjective experience.

4.4 Lateralized Response to Stress

Recent findings indicate that the right hemisphere is more vulnerable to traumatic influences than the left (Henry 1993, 1997). Reason for that is likely increased right hemispheric connection with the limbic system in comparison with the left hemisphere. The right (more often non-dominant) hemisphere is also more connected with autonomic nervous system and has predominant role in the physiological and cognitive aspects of emotional processing and is more than the left specialized for neuroendocrine and autonomic activation, for the secretion of the stress hormones, corticotrophin releasing factors and cortisol (Spence et al. 1996; Sullivan and Gratton 1999a, b; Wittling and Pfluger 1990; Schore 2001, 2002). Evidence for this lateralization provide studies dealing with the relationship between conditioned fear response and amygdala function, which show that this activation is right hemisphere dominant (LaBar et al. 1998). Also has been reported that partial kindling of the right and not the left amygdala induces a long-lasting increase in anxiety-like behavior (Adamec 1997, 1999), and that the kindling in the right amygdala induces increased production of the corticotrophin releasing factors (Adamec and McKay 1993). Recent evidence also suggests that the right amygdala is more involved in the storage of fearful faces and in the expression of emotionally influenced memory of aversive experiences in comparison to the left (Morris et al. 1999; Isenberg et al. 1999; Coleman-Mensches and McGaugh 1995; Schore 2002).

Lateralized regulation of stress responses and emotionality-related processes indicating close relationship between stress and right brain mechanisms also occur at the level of the medial Prefrontal Cortex (mPFC) (Sullivan and Gratton 1999a, b, 2002). In addition, prelimbic and infralimbic regions of mPFC have an influence to visceral motor regions, autonomic functions and emotional expression, and represent an important region for the integration of neuroendocrine and autonomic activity with the behavioral states and cognitive processes (Sullivan and Gratton 1999a, b, 2002). These studies also suggest that although the right mPFC is necessary for a normal stress response and adaptation, excessive activity of this region is predominantly maladaptive.

Influences of stress and dissociation on information transfer across the corpus callosum have also been observed (Spitzer et al. 2004). This interhemispheric dysfunction might represent a relationship between traumatic dissociation and "functional dissociation" of the hemispheres. This functional dissociation according to literature may be a form of reversible blocking of information transfer across corpus callosum (Bogen and Bogen 1969). This might explain why certain dissociative symptoms are similar to symptoms in the patients with split brain as a consequence of anatomical "dissociation" between hemispheres which occur after surgical cut of corpus callosum (Ahern et al. 1993; Galin 1974; Brende 1984; Bob 2003b; Bogen and Bogen 1969; Spitzer et al. 2004). The functional dissociation might be a defense mechanism that enables to "health" hemisphere to inhibit the negative impulses from the dysfunctional hemisphere, similarly as in psychological dissociation that inhibits a certain negative psychological impulses, which does not fit into current cognitive scheme. In this context, Nasrallah (1985) suggested that one of the vital components of the interhemispheric integration is inhibitory influence of the verbally expressive hemispheric consciousness (predominantly the left) on thoughts, intentions, and feelings from the other hemisphere.

4.5 Stress and Memory Consolidation

According to recent findings there is evidence that stress disrupts normal activity and memory consolidation in the hippocampus and prefrontal cortex (Diamond and Rose 1994; Ruel and de Kloet 1985; Payne et al. 2006). This process leads to memories that are stored without a contextual or spatiotemporal frame and produces memories that are often fragmentary, temporally and spatially disorganized, mainly because they originate from entirely unrelated events (Nadel and Jacobs 1998). Disturbed temporal memory related to stress conditions is evident from studies focused on episodic and autobiographical memories, which show temporal disorganization, fragmentation and incompleteness but not necessary absence of emotion or dissociative amnesia (Kenardy et al. 2007; Payne et al. 2006; Brewin 2007). According to these findings stress may influence atypical consolidation of short-term memory into long-term memory (Nadel and Jacobs 1998; McGaugh 2000; Debiec and Altemus 2006; Debiec and LeDoux 2006) and may cause dissociation of memory systems concerned with encoding emotion and context that occurs at psychological, physiological, and anatomical levels (LeDoux 1992, 1993, 1994; Nadel and Jacobs 1998; Phillips and LeDoux 1992; Bechara et al. 1995).

The process of memory formation is on a molecular level linked to protein synthesis that requires involvement of transcription factor CREB, expression of the brain derived neurotrophic factor (BDNF), and other molecular processes that participate in global processes of network consolidation mainly in the hippocampus and also in other brain structures (Debiec and Altemus 2006; Nadel and Jacobs 1998; Debiec et al. 2006; Lee et al. 2004). Recent molecular genetic and behavioral studies also demonstrate that spatial and contextual types of hippocampus-dependent formation of long term memory require different signaling molecules implicating distinct types of hippocampus-dependent long term memory that differ in their underlying molecular mechanisms (Mizuno and Giese 2005). Important molecule presents transcription factor CREB which is a protein that binds to specific DNA sequences and controls transcription of genetic information from DNA to mRNA in processes that enable neuronal plasticity and long-term memory formation in the brain (Silva et al. 1998). CREB specifically influences BDNF expression which as a polypeptide growth factor influences differentiation and survival of neurons in the nervous system and plays an important role in regulating synaptic plasticity, connectivity and memory storage in the CNS (Bath and Lee 2006; Bramham and Messaoudi 2005). Interesting recent findings indicate that this relationship between BDNF and cognitive processes is also significantly influenced by stress (Savitz et al. 2006; Roth and Sweatt 2011) and that especially chronic stress has inhibitory influence on BDNF synthesis, synaptic plasticity, transmission and memory formation especially in the hippocampus and neocortex (Binder and Scharfman 2004; Thomas and Davies 2005; Savitz et al. 2006; Calabrese et al. 2009).

According to recent evidence, stress influences BDNF and other molecular processes that specifically affect fixation of new information in the process of long-term potentiation and may cause spatio-temporal fragmentation of memories which is typical for dissociative states (Nadel and Jacobs 1998; Binder and Scharfman 2004; Thomas and Davies 2005; Savitz et al. 2006; Payne et al. 2006). Current findings show that extremely negative emotional experience during traumatic events or inescapable stress likely may block process of long-term potentiation in the hippocampus and medial prefrontal cortex (PFC). This disturbed long-term potentiation may influence atypical memory consolidation and memory distorsions that consequently produce consolidation process predominantly on implicit (subliminal) level in the amygdala. This blocking of higher cognitive processes mediated by hippocampus and PFC generates various unconscious automatic responses that mainly depend on the amygdala and other subcortical structures (Bob 2007; Debiec and Altemus 2006; Nadel and Jacobs 1998; Payne et al. 2006; Vermetten and Douglas 2004; Maroun and Richter-Levin 2003).

These findings are also in agreement with neuroimaging data which suggest that characteristic changes in the perfusion of limbic brain structures, such as the amygdala and the hippocampus, coincide with the high arousal and/or anxiety during traumatic recall (Vermetten and Douglas 2004). These findings suggest that a particular role in specific formation of dissociative states has the amygdala that participates in modulation of memory consolidation and also has a specific role in consolidation of the traumatic memory (Cahill 1997; Cahill and McGaugh 1998; Bob 2007; Payne et al. 2006). Typical feature of traumatic memories is that they are not acceptable for conscious awareness because of coupled strong negative emotions

(Bob 2007; Nadel and Jacobs 1998; Payne et al. 2006). High anxiety and arousal related to traumatic experience are thought to extremely focus the attention and this attentional shift may produce fragmented memories, personality fragmentation (Bob 2007; Vermetten and Douglas 2004), psychological automatisms and lowering of mental level as described by Janet (Bob 2003a, 2008; Ellenberger 1970; Frankel 1996; Havens 1966; Putnam 1997; Van Der Hart et al. 2005).

In general these influences related to stress and dissociation may create specific state dependent memories based on a principle that something learned in one neuropsychophysiological state is best recalled in the same state (Brown 1984). In this context, process of dissociation and changes in memory consolidation may be understood within the framework of state dependent learning that was originally studied in experiments with animals under influence of drugs, which typically influence memory formation and a neural state that is not possible to recall without the drug. Similar experiments in humans also confirm a relationship between a type of the task and its recall (Brown 1984). These effects were also found in sleep and circadian rhytmicity (Holloway 1978), postictal states (Overton 1978), as a consequence of changes in mood (Bower 1981), in cases of maniodepresive states induced by dextroamphetamin application (Henry et al. 1973), in stress situations or in emotional states (Henry et al. 1973; Pearce et al. 1990), and also using hypnosis (Brown 1984; Bower 1981) that also presents specific form of dissociative state (Hilgard 1986).

These data also implicate important consequences for psychotherapy that in its effects is based on memory reconsolidation in safe and non-threatening conditions, which leads to neurobiological reprocessing of memory traces. This reconditioning and reconsolidation is therefore possible only by re-experiencing of the traumatic memory in a new and safe situation, for example during psychotherapeutic process which enables integration of the dissociative state (Bob 2007). During this process implicitly consolidated traumatic memory in subcortical structures, mainly in the amygdala, is probably transformed from automatic into higher levels of conscious experience by long-term potentiation in the higher-level structures of the CNS such as medial PFC and hippocampus. In this context memory reconsolidation likely represents a process that enables successful transformation of dissociated, automatic and implicitly consolidated traumatic memories, located mainly in the amygdala, to higher levels of conscious experience in the higher-level structures such as medial PFC and hippocampus. In addition, this memory reconsolidation is again related to BDNF synthesis and other molecular processes that influence modification of synaptic plasticity, transmission and new memory formation especially in the hippocampus and PFC, which include gene expression that is required for hippocampus-dependent long term memory formation (Debiec et al. 2002; Morris et al. 2003; Mizuno and Giese 2005; Savitz et al. 2006). This new gene expression and protein synthesis is based on endogenous BDNF-TrkB signaling in synaptic consolidation by long-term potentiation (LTP) that enables immediate early gene Arc activity (activity-regulated cytoskeleton-associated protein) (Bath and Lee 2006; Bramham and Messaoudi 2005; Soule et al. 2006). Important factor in this new gene expression is also the transcription factor CREB, which is required for hippocampus-dependent long term memory formation (Nadel 1994; Mizuno and Giese 2005). The memory related new expression through transcription factor CREB is activated by signaling pathways that include Ca(2+)/calmodulin kinases (CaMKs), protein kinase A (PKA) and the mitogen activated protein/extracellular signal-regulated kinases (MAPK or ERKs) (Mizuno and Giese 2005; Rattiner et al. 2005).

These findings indicate that psychotherapy as a form of memory reconsolidation may specifically influence and modify brain functions, metabolism in specific brain structures and also genetic processes (Kandel 1998, 1999; Gabbard 2000, 2007). This view corresponds to Janet's definition of dissociative state as an automatic process which does not fit into current cognitive scheme and without successful reprocessing (or reconsolidation) remains dissociated also during recall of dissociative state because of the specific neural substrate of dissociated memories (Bob 2007). Similarly, Putnam proposed that treatment of dissociative disorders and post-traumatic stress disorder is based primarily on psychotherapeutic, cognitive and behavioral modification approaches focused on "detoxifying" traumatic memories (Putnam 1995).

Based on these findings disunity of conscious experience leading to splitting or dissociation of consciousness is closely linked to disunity on neural levels that lead to various neuroendocrine, immune and metabolic disturbances that increase entropy of the mind and brain, and produce disorder in the organism per se.

Chapter 5
Freud's Ωmega and Self-Reference

A basic principle of mind functioning is that it cannot be fully integrated and must contain a free energy between discrete mental states that through differences between them enable recognition and awareness of the external and internal world. This form of internal disunity of the mind and brain likely presents basic 'code' that defines relatively stable differences and enables their recognition in mental and physical space. This basic aspect of using "free energy" enables recognition and this psychic energy is "bound" to contextual and emotional value of a stimulus 'meaning' and is linked to specific neural patterns related to mental and behavioral activities. In the case that recognition leads to interference, which means that recognized object is related to controversial contextual and emotional meanings, for example love and hate with respect to the same persons, it results to intrapsychic conflict.

In this context, Freud (1895) in his "Project" discussed that psychological disunity arises during creating of a judgment which may cause divergent components that disturb complex thought processes. Freud also suggested that psychological disunity is represented by a certain form of "neural disunity". This issue in principle presents a basic problem for neural sciences how the *psychological disunity* in the *neural system* may be *represented.*

5.1 Freud's Concept of Omega Neurons and Problem of Qualia

In the light of Freud's thoughts that neural disunity is related to intrapsychic conflict, current findings suggest that mental disunity is likely specifically presented on a neural level as disrupted organization in neural communication and binding which disturbs global organization of information processes related to conscious disintegration and repression. In this context, dissociation likely is related to conflicting stimuli and pain experience per se (Villemure and Bushnell 2002; Bob 2008; Craw-

P. Bob, *The Brain and Conscious Unity,* DOI 10.1007/978-1-4939-2700-5_5

ford 1994; Eccleston and Crombez 1999; Feldman 2004) which lead to alterations in "self-representation" and specific changes in binding of various parts of the brain which define dominant interpreter's access and contents of awareness (Baars 1988, 2002; Kanwisher 2001; Bob 2008, 2011). On the other hand in agreement with general thermodynamic and neuropsychological principles Freud (1895) proposed that integrated state of consciousness without a conflict might be linked to "neural unity".

In this context, Freud in his Project also suggested a basic scheme that observable psychological phenomena depend on the neural structures of the brain based on the three systems of neurons phi (φ), psi (ψ) and omega (ω). According to this concept omega neurons represent ultimate instance of psychological and perceptual reality linked to awareness, self-reflection and self-reference. The system of omega neurons is excited along with perception and states of excitation of these neurons give rise to various qualities corresponding to conscious sensations. Freud also thought that our consciousness furnishes only qualities, whereas science recognizes only quantities and the nervous system transforms external quantity into quality. He postulated that omega neurons are influenced by smaller quantities of energy and behave like organs of perception. According to this concept, Omega neurones are incapable of receiving "quantitative excitation" (Qη) but they are filled with the minimum of Qη which is the fundamental basis of consciousness. Freud also proposed that the sense-organs allow the stimulus transfer which as perceived difference is transferred to phi-neurons. Then the signal proceeds to omega via psi where it is experienced as conscious sensations of qualities.

According to Freud's proposal phenomena of consciousness are linked to excitatory processes in the omega-neurons that represent the subjective side of all psychical events and are inseparable from the physiological mental process as a subjective side of the physical processes in the nervous system (Freud 1895). Omega neurones according to Freud's thoughts must have a discharge, however small, and there must be a way of filling them with Qη in the small quota required. But as Freud wrote it is not possible to suggest what the original biological value of the omega neurons was (ibid.).

From these ideas written in Freud's "Project" it is evident that Freud anticipated recent problems of neural mechanisms related to "Qualia". Qualia represents a basic problem in the philosophy of mind that refer to the introspectively accessible, phenomenal aspects of mental experience that most likely represent irreducible and non-physical entities (Dennett 1990, 1991; Schrödinger 1944). For example, Ramachandran and Hirstein (1997) proposed basic principles of qualia— The first is that Qualia are irrevocable and indubitable, which means that they present specific experience, i.e. red is red and not green. The second is that once the representation is created, what can be done with it is open-ended, which means that various associations may be related to this percept (for example red apple produces associations that it is possible to eat it or give to someone and so on). The third principle is short-term memory that creates a representation of the input. Additional, the fourth principle links qualia and attention, and represents specific selection process which is necessary to create sensory representation of the stimulus (Ramachandran and Hubbard 2001).

The problem of qualia has been also discussed in psychoanalytical literature and represents a controversial problem that may lead to neural reductionism (Pulver 2003; Vivona 2009; Talvitie and Ihanus 2011). As mentioned before the possible solution of the problem might provide general principles of isomorphism used in mathematics (Marker 2002; Mazur 2008), which state that two objects are isomorphic when any property described in scientific description of one phenomenon is on a different "isomorphic" level valid for the objects in another descriptive domain. Applied to experimental work in science isomorphic description is focused on analogies between the two related fields based on "superstructure", which underlies both of them and represents their conceptual unity in duality.

In this context Freud's postulate of omega neurons represents a certain kind of "superstructure" that has specific quality completely different from neurons phi and psi. In fact "they are" and "they are not" just neurons, because they produce qualia which according to current neuroscience may not be directly provided by a specific neuronal group or structure and specific mechanisms of information integration in the brain (the so-called "binding problem"). Based on principles of qualia "omega neurons" have attentional functions and in fact they provide contextual processing that determine what details of the processed information will actually be presented in conscious awareness. Consequently, these contextual mechanisms of attentional sensitivity may influence processing of various aspects of perceptual information and enable to discover events that in certain contextual frameworks are presented as potential and subthreshold with respect to actual conscious experience.

5.2 Concept of Omega Neurons and Modern Chaos Theory

During recognition and selective attention processing, dynamical interactions of various independent activities produce neural "recognition patterns" that enable to compare sensory information with patterns in memory and distinguish differences between these patterns. In this context, attentional selection means comparison and selection of a perceptual information during process of 'interactive competition' (Desimone and Duncan 1995). This process likely is based on repeated comparisons of various competitive neural patterns representing possible interpretations of the received information during selective attentional processing (Desimone and Duncan 1995; Baars 1988, 2002; Kanwisher 2001; Bob 2011). As a consequence, the various properties of an object represented in distinct cortical regions enhance each other and suppress neural representations of competing objects. Competitive interactions across cortical areas result in domination of perceptual representations by properties of a single object. Selected interpretation from this competition and its neural pattern represent the output of the recognition process, which is subjectively experienced as mental representation of the interpretation.

With respect to recent evidence about attentional processing these complex dynamical interconnections forming various neural patterns and also comparisons among them produce "differences" that define basic "code" for recognition which in every comparison during attentional processing means one bit of information and on highest levels of this hierarchical process of recognition it is one bit of information that presents a final step of the recognition process of which output is yes or no (1 or 0 which is one bit of information). In this context, interesting physical quality of the "omega neurons" that Freud postulated is their very low energy of excitation. This postulate from the current point of view might be reasonable because energy which is necessary for recognition of one bit of information in principle might be very low energy at about 10^{-21} J (according to Landauer's principle *k T log2*; Chap. 1.3.1.2.). This very low energy means that the process of recognition in its final steps might be processed on very sensitive microscopic level.

Based on these postulated qualities omega neurons in principle might be very sensitive "chaotic" devices that using very small and sensitive discharges during and after recognition process enable to produce massive neural responses in other neurons which consequently produce motor and behavioral responses (i.e. in principle omega neurons might use chaotic "butterfly effect" and through very small changes to produce highly significant effects on macroscopic scales). This conceptualization of omega neurons in its basic principle that Freud proposed is in agreement with current chaos and complexity theory of the brain functions in which sensitivity during "critical times" significantly influences rapid shifts in the system's evolution which later may emerge as very different macroscopic patterns of neural activity and mental processes (Elbert et al 1994; Freeman 1983, 1991, 2000; Birbaumer et al. 1995; Kantz and Schreiber 1997; Meyer-Lindenberg et al. 2002; van Putten and Stam 2001; Faure and Korn 2001; Globus and Arpaia 1994; Korn and Faure 2003).

Because omega neurons were postulated to explain and enable qualia and integrated conscious experience most likely their "dysfunction" could be linked to intrapsychic conflict and conflict monitoring processes which lead to splitting (or dissociation). Because mind needs to have permanent intentional focus, it tends to dissociate due to contradictory tendencies related to a conflict. Due to the conflict, process of splitting is related to active formation of mental associations and complexes that although they represent organized structure, contain contradictory tendencies which cannot be connected and put together (e.g. ambivalence). These contradictory tendencies produce processes of internal disunity and psychological automatisms related to unconscious mental activity. These conflicting contradictory tendencies are represented by mental events that do not fit into predominant cognitive scheme and cannot be integrated into awareness.

As a possible mechanism that brain and mind may use in conflicting situations, Freud (1930) proposed process of sublimation that enables to redirect free energy of innate mental patterns and their conditional forms, and transform it into novel conscious intentions. In this context, sublimation means transformation of usually preferred innate or learned patterns of mental organization that tend to create typical neural and behavioral patterns into new and unusual forms. In this context, Freud's concept of sublimation also corresponds to current views about neural chaos that enables to create novel forms of mental activity and behavioral patterns. For example, Freeman proposed that chaos could underly basic forms of collective neural activity in perceptual processing including ability to access memorized sensory patterns and learning novel sensory information (Freeman 1991, 2000, 2001; Skarda and Freeman 1987). Freeman also proposed that chaos may explain brain ability to respond flexibly to the outside world and to generate novel activity patterns that are subjectively experienced as "novel" ideas and intentional behavior (Freeman 1991, 2000, 2001; Skarda and Freeman 1987). In this context, chaos theory enables to understand collective neural activity and brain functions as a global integrative process generated in the network of synaptic connections representing behavioral "intentional" patterns that may be modified by learning (Skarda and Freeman 1987; Freeman 2000).

Within this framework Freeman (1999) proposed that the linear view of the stimulus-response reflex determinism just partially may serve as an appropriate concept and explanatory principle for the behavioral dynamics and suggested that behavioral responses and intentional behavior may be better understood as consequences of nonlinear chains of various stimuli and responses. Freeman (1999, 2000) also suggested that chaotic and complex self-organization of multilevel interactions between microscopic neurons in assemblies and the macroscopic emergent states is not possible within the concept of "linear causality", and must be replaced by "circular causality" (or reciprocal causality) that enables to reflect extensive relations of mutual dependencies, actions and influences. These nonlinear chaotic processes represent a consequence of high system complexity, when the system involves a large number of complex interlinked and simultaneously active neural assemblies and runs in a desynchronized parallel distributed mode which can lead to self-organization (Freeman 1991, 2000, 2001; Velazquez et al. 2003) and typical dynamical instabilities in mental phenomena (Atmanspacher and Fach 2005). In this sense the intuitive concept of omega neurons and process of sublimation are closely linked to basic findings in modern neuroscience.

5.3 Freud's Basic Principles and Physical Theory of Observation

Although Freud's basic principles of mind functioning are intuitively quite close to recent complexity and chaos theory of the brain, at the time of the "Project" basic paradigm of thermodynamics was linked to general physical concepts within the framework of Newton's theory of space and time. This paradigm was basically linked to linear view of causality and the idea of basic structural units that may be composed and enable to create structures similarly like stones may create buildings. Within this paradigm Isaac Newton proposed mathematical and physical principles of existence of infinitely small entities that he called "fluxions" that may be put together and create finite objects or processes in the universe. In this sense, very small entities like atoms and other elementary particles that in principle may be considered as infinitely small building blocks that using differential and integral analytical calculus ("multiplication") of these infinitely small fluxions may provide finite result (Whiteside 1967; Newton 1686/1974, p. XVIII). For example summation of infinitely small volumes may provide finite volume of a body or summation of infinitely small distances may provide finite distance between some places. On physical level of the universe this theoretical concept led to postulation of basic elementary "particles" i.e. "fluxions" of "atomic" undividable nature that, as basic structural elements from which all objects in the world are composed, create the universe. These particles as basic structural elements also provide basic reference for all physical processes in the form of absolute space which is linked to gravitation force among all material objects in the space with Euclidean geometry.

In Freud's time, Newtonian physics was taken as a reference frame for understanding of all physical, biological and psychological processes. But some basic conceptual differences in comparison to classical physics that is possible to find in Freud's theory were approximately at the same time proposed in physics. This completely different view of space and time and geometry of the universe discovered Riemann and Einstein (Riemann 1873; Einstein 1916/2005; Monastyrsky 1999; Derbyshire 2004). As an extension of Euclidean "flat" space, Riemann described curved space with the so-called "Riemannian metrics" (Monastyrsky 1999; Plotnitsky 2009). The level of curvature in this space metrics characterizes a difference between usual "flat" Euclidean space and the "Riemannian space". Intriguing and groundbreaking idea how to apply the Riemannian metrics was published by Einstein in 1915, who found that gravitation may be more generally described as a specific change of spatial structure caused by geometrical curvature that changes structure of the space and time, for example in extreme cases "black holes" have extreme curvature leading to distortion of the space and time and cumulation of extreme mass, energy and gravitation (Einstein 1916/2005; Laue 1950; Pauli 1958; Beiser 1995, 1916/2005; Thorne 1994; Carmeli 1999; Field 1999). This geometrical approach included in the relativity theory enables to describe different observers' viewpoints leading to different measures as typical changes of space and time metric, for example length between two points may be measured differently by two different observers (Einstein 1916/2005; Penrose 2004).

In this context, it is a general law of nature that reality most likely is not objective per se and there are not independent observers, who could know true about the Universe as it was proposed in Newtonian classical physics. According to these findings every observer has a specific viewpoint which influences observation process and also observed reality and its measures. On most general level this principle described Kurt Gödel in the well known "Gödel's incompleteness theorem" which states that any mathematical axiomatic system using its the regular deductive rules contains at least one statement which is within the framework of this axiomatic system true and false at the same time (Boyer 1968; Chaitin 1982; Rucker 1982; Penrose 1994; Hofstadter 1999; Nagel and Newman 2002). As a consequence it means that likely it is not possible to find rigorous system of mathematical truths that would be without any paradox and conflict in observation processes, which implicates the same consequences also for physical theories and scientific knowledge in general. For example, from this point view most likely it is not possible to find theory of everything based on mathematical axioms that could explain the universe as a whole.

On this general level of thinking the same "objective" situation may be understood differently from the point of view of different observers which is especially true for psychology where using transference each individual creates unique intentional space and time. At this point Freud's concept of scientific psychology used different paradigms than was usually applied in biological and psychological sciences at the end of nineteenth century.

5.4 Brain Observers and Freud's Concept of Omega Neurons

Analogical problem of observation that plays a role in "reality formation" and brain processes explicitly emerges in the quantum theory. Heisenberg (1958, p. 22) thought that the transition from the "possible" to the "actual" in quantum theory takes place during the act of observation (1958, p. 81), which was shortly summarized by Bohr (1958, p. 81) that: "in the great drama of existence we ourselves are both actors and spectators." Similar view proposed also von Neumann (1955, p. 418), who emphasized that interpretation of the quantum theory requires the existence of a subjective (i.e., conscious) observer, who's mental activity influences physical processes. In this context, von Neumann divided the world into three parts (I, II, and III). According to this division part I is everything up to the retina of the observer, part II includes observer's retina, nerve tracts and the brain, and part III is observer's abstract 'ego' (Neumann 1955, p. 420). In addition, von Neumann also considered the case when the observer forgets the information and found that this process also leads to an increase of entropy (Neumann 1955, pp. 417–437).

Later Bass (1975) suggested that during the act of observation the entry of information into the consciousness of an *"observer" inside the brain* influences the observer's central nervous system and trigger causal mechanisms of neural response that consequently may induce muscular movements and the response. In this context Bass (1975) proposed a semirealistic neurochemical model that describes the entry of a datum into the consciousness of an observer that influences excitation of a nerve cell in the observer's central nervous system. He proposed that mind can induce muscular movements by choosing to note or not to note a relevant datum originating from specialized elements of the nervous system. Similar view developed also Mould (1995, 1998), who proposed that the conscious brains have "inside observers" in contrast to a conscious observer existing outside the system, and suggested that there exists a neurological mechanism responding to the presence of an "inside observer" with experimentally testable consequences. The mind-brain interaction during act of observation studied also Stapp (2001), who proposed that attentional mechanism related to observer's mental activity may present a phenomenon analogical to the so called quantum Zeno effect. Zeno effect is similar to the Zeno's paradox about impossibility of motion (for example to go from A to B means that first step will be half of this distance then half of the half, i.e quarter of the distance, then half of the quarter,... etc.). Zeno effect is analogical to a situation when someone says something but he/she is repeatedly disturbed during talking and starts again and again the same. In this context, Zeno effect can be defined as a phenomenon in which time evolution of a quantum object can be suppressed by measurement and because of that the system repeatedly returns to its initial state because of repeated measurement which does not allow to the system to finish its evolution and forces it to start its evolution again, for example in decay of unstable particles and other experimental arrangements (Sudarshan and Misra 1977; Igamberdiev 2007; Maniscalco et al. 2008). Stapp proposed that the Zeno effect during brain processing could explain several phenomena related to keeping of the focused attention on a task. In line with psychological data (Pashler 1998; Stapp 2001) a time between 100 and 300 msec seems to be needed to fix attention without its interruption and for example increased effort that interrupts attentional focus may start the process again (Stapp 2001, 2004, 2005; Schwartz et al. 2005).

Quantum theory also defines that quantum systems may include various spatially distant objects and when an observation or measurement is performed on one of the objects which is a part of the system also other objects that are parts of the system will respond immediately on a principle of instantaneous action (Einstein et al. 1935; Bell 1987). It results to paradox, where the measurement of a physical quantity in one subsystem affects the measurement of a physical quantity on another subsystem in the spatially separated system, the so-called quantum non-locality. This immediate action on a distance contradicts the theory of relativity which defines that limiting speed of any signal connecting two events in the space is the velocity of light. A historical origin of the term "quantum non-locality" is the so-called Einstein–Podolsky–Rosen paradox (EPR paradox) (Einstein et al. 1935; d'Espagnat 1976; Bell 1987; Penrose 1994; Wheeler and Zurek 1983) and this principle was experimentally confirmed on pairs of photons (Aspect et al. 1982; Wheeler and Zurek 1983; Laurikainen 1988; Penrose 1994; Aspect 1999; Schlosshauer 2005). For example, pair of two connected spatially distant particles in the so-called non-local (or entangled) state can be used as an information channel that can enable to transmit the information instantaneously, when an influence on the first particle immediately influences the second one, independently of the the distance between them. This principle is used in the modern theory of quantum computation that enables to perform computational and cryptographic tasks which is not possible in the classical non-quantum systems (d'Espagnat 1976; Bell 1987; Penrose 1994; Nielsen and Chuang 2000). Although this information is intuitively controversial the study of the information-processing in quantum systems is a subject of the quantum computation theory that has a specific aim to develop quantum computers (Haroche and Raimond 1996; Nielsen and Chuang 2000).

In principle, this process of quantum instantaneous action among subsystems of the quantum system could explain synchronous immediate connections in the brain but neurobiological mechanisms of this process are not known although various concepts and hypotheses have been proposed (Marshall 1989; Beck and Eccles 1992; Crick 1994; Crick and Clark 1994; Woolf and Hameroff 2001; Fröhlich 1968, 1970, 1975; Smolin 2004; Karolyhazy 1966; Ghirardi et al. 1990; Hagan et al. 2002; Hameroff and Penrose 1995, 1996; Penrose and Hameroff 1996; Zaccai et al. 1998). Because consciousness and process of observation is an important part of the physical reality, in principle description of the observer's participation in measurement process may play an important role for explanation of a specific relationship between quantum physics and general relativity theory which both include specific influence of the observer in reality formation (Kragh 1999; Callender and Huggett 2001; Penrose 1989, 1994, 2001, 2002; Tegmark 2000).

In this context, relativity theory and quantum theory brought completely novel view of physical processes which based on correspondence principle as proposed by Bohr (1958) means that relativity theory and quantum physics both represent more general theories that using certain limitations are able to provide the same description like classical Newtonian physics which is especially clear in the relationship between Newtonian mechanics and Einstein's special relativity theory. But the correspondence principle may also mean that certain phenomena that were not included in Newtonian classical mechanics could exist in macroscopic dimensions of the universe. There is evidence that at least two macroscopic phenomena exist on which it is possible to describe influence of "observation process" on observed and measured phenomena in nature. These are Light and Mind. Due to the specific observation process related to experimental setup, light may be observed as the wave and/or as moving particles (photons), and similarly influence of observation is specific in contextual interpretation during brain information processing of sensory stimuli (Desimone and Duncan 1995; Baars 1988, 2002; Kanwisher 2001; Bob 2011).

As a consequence it means that the observation process in principle is defined by its viewpoint which may be determined by various physical and biological characteristics that may influence state of the observer. These characteristics are context specific and depend on definition of the observer and the system in which the observing process exists. This means in general, that any physical and biological process may be specifically defined and influenced by its observation, i.e. state of the observer and the observer's relationship with respect to the observed event or process.

As for the brain inside and its internal observer, this general physical principle is in agreement with neuroscientific evidence that different "inside observers" produce different "views" of processed information. In the brain these observers are related to specific forms of attentional processing that produce different forms of observation of the brain information which is linked to specific forms of binding and complexity of the various aspects of observed information which is specifically influenced by physiological state of the brain and attentional focus related to observer's conscious awareness (Desimone and Duncan 1995; Baars 1988, 2002; Kanwisher 2001; Bob 2011).

For example, Petitot (2003) in his formulation of "neurogeometry" suggested interesting theoretical concept that based on mathematical principles of the relativity theory could explain the problem how the internal geometry of the visual cortex during vision can produce the geometry of the external space. Petitot has found that the specific mathematical condition of integrability reflecting connectivity of the various parts of the brain represents neurogeometrical condition of binding mechanism among distributed neurons that allows understanding of typical Gestalt phenomena. Petitot also proposed that a synchronized wave of activity propagating in the functional architecture is geometrically equivalent to the integration of a specific differential equation describing the internal brain processes using Riemannian geometry used in general relativity theory which perceptively enables to understand problem of binding as a physical-geometrical process (Petitot 2003). Similar geometric interpretation using Riemannian geometry for description of the brain developed also Pellionisz and Llinas (1982, 1985) and Llinas (2002), who studied Riemannian representational space creating dynamic correspondence between events in the external world and their specification in the internal spacetime geometry. They used this description of external space-time events to characterize various reference frames in the internal space and time information processing.

Application of this geometrical concept to the brain may implicate that brain metrics may change due to a specific point of view of the "internal" observer or "brain interpreter" during attentional processing that is linked to conscious awareness. From this point of view the binding problem may be resolved by a non-existent distance among the parts of the brain from the point of view of the brain's interpreter, because of the different space metrics in the neural space that may enable brain synchronization and coherence (Bob 2009, 2011).

The different perspectives between external and internal observation may play a role, and may imply that the spatial distance between two points in the brain may be different from the viewpoint of an external observer in comparison to an internal observer, which is the brain interpreter or observing mind itself. In this context, a change in the brain geometry could play a fundamental role in brain functions during very short periods of its functioning (in order of miliseconds or less) in which attention is fully focused on internal processes and is closed to external stimuli. Those very short periods of time during discontinuous flow of "flashing" attention may cause that the brain internal geometry is closed with respect to outside stimuli which could correspond to "closed" Riemaniann metrics, which is related to the length contraction in the closed space that could enable neural binding with "zero distance" among simultaneously active groups of neurons from the point of view of the brain interpreter.

From this point of view the states of consciousness may be explained by different levels of binding and information integration related to different and specific forms of focused attention that defines "the brain observer" during information process. This specific point of view produces observer specific contextual binding of distributed information in the brain related to simultaneously active neuronal assemblies bound into the specific coherent whole. Typical examples how the various brain observers may differently interpret the same stimulus provide ambiguous figures (Fig. 2.1.), or for example Rorschach plates or pictures used for interpretation in Thematic Aperception Test, which reflect interpretation and contextual cognitive

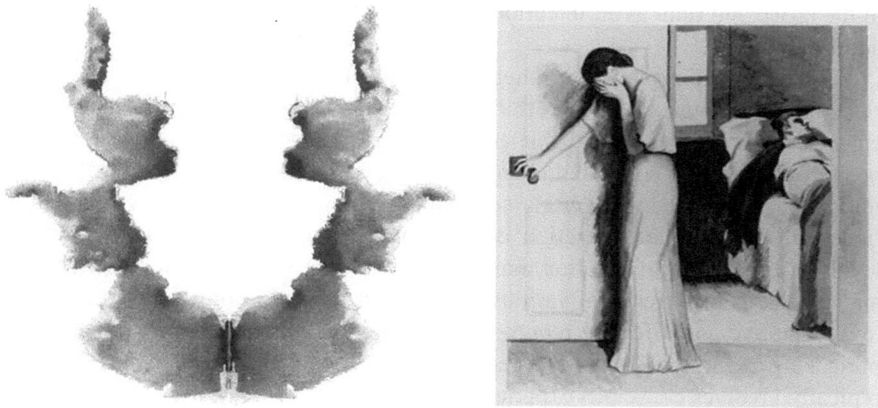

Fig. 5.1 Examples of Rorschach test (*left*) and Thematic Apperception Test (*right*) that represent typical stimuli eleciting multidimensional meanings and interpretations

formation of perceptual reality as it occurs during brain information processing (for example Fig. 5.1.).

Other examples of attentional specific processing corresponding to different "observers" in the brain present various forms of attentional or emotional modulation for example related to experience of pain, which means that in certain mental state observer feels pain and in another mental state does not feel the pain due to the modulation effects for example ín hypnosis (Hilgard 1986; Bob 2008). Similarly hypnotic modulation of the observer's conscious mind is able to bring to conscious awareness various dissociated memories which are not available in normal state of consciousness and hypnosis also may change subjective length of experienced time or cause various perceptual distortions (Hilgard 1986; Bob 2008, Noreika et al. 2012; Bowers and Brenneman 1979). Other observer specific changes of mental state show cases of dissociative identity disorder in which different "observer specific" EEG patterns have been documented (Schenk and Bear 1981; Mesulam 1981; Coons et al. 1982; Benson et al. 1986; Spiegel 1991; Perrine 1991; Hersch et al. 2002).

In this overall context of modern scientific findings, Freud's view in which omega neurons produce a specific act of observation that is related to excitation of psychic energy that enable to bind the free energy and create person-specific mental interpretation of the world is in agreement with current physical theories suggesting that there are no independent observers and all observation processes are related to observer-specific descriptions of space and time. This Freud's view of consciousness is not "objective" in the sense that conscious mind would be just epiphenomenon of interacting particles creating living bodies and similarly like in modern physical theories the concept of consciousness in Freud's theory attributed to "omega neurons" activity is closely linked to specific process of observation in which the 'free energy' related to "cathexis" in omega neurons is bound and occurs in the space and time as an intentional event related to interpretation, expectations and finality.

With the aim to explain how omega neurons work during events that create experience of consciousness, Freud (1895) used the term "cathexis" (in German Besetzung) as an investment of psychic energy into thoughts and expectations that have positive or negative emotional charge and create psychological "complex", which links thoughts related to a certain contextual framework with intentions and emotions. This specific interpretation is based on selection from various hierarchical levels of processed information that enables to create various contextual schemes and conscious experience of various details of processed information. In this context, cognitive and affective representations of one's identity or the subject of experience present a basis for self-recognition as a specific cognitive process that typically involves conscious experience and interpretation activity.

Disruptions of these self-interpretation processes likely represent a neurophysiological substrate for the process of fragmentation of consciousness due to cognitive interference (Bob 2008, 2011). This process is subjectively experienced as conflicting and in agreement with Freud's concept increases "neuronal energy" and leads to a need to find a new cognitive scheme that enables through new "cathexis" and "sublimation" to integrate dissociated memories and perform memory reconsolidation which significantly influences the brain and body on various functional levels. In this sense, dissociation (or splitting), in principle, may be explained by various levels of disturbed binding that negatively influence patterns of connectivity, complexity and synchronous activity constituting adaptive integrative functions of consciousness (Nadel and Jacobs 1998; Tononi et al. 1998; Payne et al. 2006; Bob 2007, 2011). On psychological level, dissociation presents a mental event in which some contextual element is conflicting and although it is a part of the mental scheme at the same time it does not fit into this contextual framework.

According to Freud's thoughts in the "Project" (1895) the experienced quality linked to contents of consciousness within the contextual system is related to investment of psychic energy (cathexis) which is mainly based on series of sensations of pleasure and unpleasure. In agreement with the 'inertia' principle, which means that a level of neuronal energy tends to move towards states with minimal energy, a typical trend in psychic life is to avoid unpleasure. This principle implicates that unpleasure presents force raising the level of "quantitative excitation" ($Q\eta$) in psi neurons which is detected by omega neurons and leads to conscious experience. The conscious experience of this tension increases the need to use psychic energy and find contextual framework of activities resulting in a pattern of behavior that is linked to discharge of omega neurons. When $Q\eta$ is too high and the tension is not adequately compensated by resulting activity producing cathexis that would enable omega neurons to discharge, then the accumulated energy in omega neurons is experienced as unpleasure. When this excitation in psi and omega neurons is excessively large it produces pain as an extreme level of unpleasant experience which is related to various forms of pathological discharges of the psychic energy leading to repression of the energy. This repression of psychic energy in principle may explain various psychopathological symptoms and disorders, and may also explain conversion of repressed psychic energy into somatic symptoms. In this context different "observers" in the brain present various forms of attentional or emotional modulation for example related to experience of pain, which means that in certain mental state observer feels pain and in another mental state does not feel the pain due to the modulation effect as for example ín hypnosis (Hilgard 1986; Rainville et al. 2002; Bob 2008) that enables modulation of the observer's conscious mind and may bring to conscious awareness various dissociated memories which are not available in other states of consciousness (Hilgard 1986; Lynn et al. 1994; Bob 2004, 2008).

These specific states of consciousness and attentional sensitivity are mediated through various levels of connectivity patterns and large scale brain information processing that enable to combine information from various sensory modalities with information in memory and create coherent contextual frames (Fries 2001; Singer 2001; Varela et al. 2001; Zeman 2001; John 2002; Crick and Koch 2003; Bob 2009).

In this context, dissociation presents a mental event in which some contextual element is conflicting and although it is a part of the mental scheme related to the event, at the same time it does not fit into this contextual framework. This process is subjectively experienced as conflicting and provokes a need to find internal integrity and symmetry, which is possible using various methods psychotherapy that enables to find a new cognitive scheme providing a possibility to integrate dissociated memory of the event and perform memory reconsolidation which significantly influences the brain and body on various functional levels (Nadel and Jacobs 1998; Payne et al. 2006; Bob 2007). These findings suggest a new paradigm for understanding of the mental world and a possibility to influence the physical body using psychotherapy through specific changes of contextual mental processing related to learning process that may specifically influence and modify brain functions, metabolism in specific brain structures and also genetic processes (Kandel 1998, 1999; Gabbard 2000, 2007).

5.5 Conscious Observers in Intentional Space and Time

According to Freud (1930) repression represents the withdrawal of this psychic energy that create space of emptiness, it may be compensated by sublimation represented by redirection of the psychic energy to another contextual system of goals, wishes, thoughts, feelings and emotions. Based on clinical evidence of sublimation in psychotherapy, but also in agreement with current scientific view related to complexity and chaos theory, dissociated mind has an existing potentiality of integrated mental states that compensate contradictory tendencies and dissociated 'free energy', and enables to 'bound'or reintegrate it into specific mental images and association patterns'. In this context, mind may use its free energy in the external world or loss the energy, but it may also integrate the 'intentional' energy, decrease entropy and increase brain/mind organization and information content. This ability of mind to integrate the intentional energy represents its unique function that enables to transcend the contradictory tendencies and "neural disunity". In this context, there is evidence that self-reflective consciousness may integrate brain functions and enable access between otherwise separated neuronal processes (Baars 1988, 2002; Kanwisher 2001; Bob 2011). This relationship likely means that highly integrated mental activity and neural binding, for example as a consequence of psychotherapy, meditation and other exceptional self-reflective experiences, implicate decreased entropy. This decreased entropy like in other thermodynamic systems tends to create higher levels of organization that in the brain may produce typical neural patterns leading to increased level of order and integrity in mental 'associative connections' producing self-reflective experience that increases information content of the mind and creates new view of the intentional space and time.

The same principles of psychological interpretation and the role of the observer are applicable also for scientific description of the world which also depends on specific types of observation and related empirical data. Interesting findings that various critical steps in science are not produced by logical thinking but intuitive change of the "world view" was introduced by Thomas Kuhn in his book "Structure of Scientific Revolutions". Based on historical data documenting evolution of scientific theories Kuhn (1970) proposed that science in its evolutionary steps has discontinuous transitions which are represented by new theories that are not deducible from previous knowledge and represent qualitatively new discovery that is characterized by its "paradigm."

Paradigm means accepted style of thinking and interpretation of scientific data in the framework which is not scientific per se. This paradigmatic framework is based on specific contextual understanding and interpretation reflecting predominant view in the society based on subjective meanings represented by religious or philosophical concepts of reality. Kuhn (1970) emphasized that due to this interpretative aspect of knowledge scientists see different things when they look from the same point in the same direction. Paradigm may be religious or atheistic, geocentric or heliocentric, classical or relativistic and so on. In fact paradigm reflects predominant interpretation of reality among others that is momentarily preferred in the society and accepted as the only one (or at least dominant) truth about life and the universe. The paradigmatic "truth" related to scientific knowledge is more prejudice than discovery and typically it is linked to unconscious postulates about nature per se. This prejudice is based on emotional meaning of the interpretation, which is closely linked to expectations that every scientific finding may provide with respect to its applications.

In this sense, the paradigm represents a complex view consisting of feelings, desires and intentions that create global picture of reality. Typical feature of any paradigm is overgeneralization, which based on known findings explains also unknown parts of reality as a particular aspect of this "knowledge" that creates a fiction that everything is known in its basic essence. Based on this collectively accepted interpretative error, paradigm manifests itself as an ownership of truth represented by shared feelings and opinions creating historically and culturally characteristic individual projections that interpret basic features of reality and from the particular knowledge it creates most general mental representational maps of the universe.

The ownership of truth is usual defense mechanism against fear of unknown which creates illusion of "total" predictability in "predictable" as well as "unpredictable (random)" universe.

In the context of Freud's theory paradigm is based on the transference which causes that we project our own mental states and intentions onto other things which enables that the free psychic energy is 'bound'. Based on this principle mind and brain temporarily identify unconscious and implicit "internal" intentional energy with real events and processes in the external world that exist as representations of the internal space of potentialities -goals, wishes and intentions- that have energy which can be bound into these intentions and create mental representational maps. In this context, paradigm in science in principle is not different than other wishes and intentions and represents collectively shared personal views and opinions. Novel and groundbreaking findings in science are therefore typically and frequently linked to very different view of the world and the universe. For example, from the point of view of Newtonian and Cartesian scheme of the world, everything is just composed from basic elements of matter. In this view of the universe consciousness is just random artifact of this composition or it may be understood as God's immaterial existence as Descartes proposed in his concept of Res Cogitans representing mental existence (in principle participating on God's essence) and Res Extensa (representing material existence in the space). On the other hand relativity theory and quantum mechanics postulate in their basic scheme observer's existence that represents fundamental condition how the physical process will be described and the observing process plays a role in the reality formation. From this point of view it looks like that every philosophy of science creating a paradigm may produce significantly different interpretations of the world and universe and even it is just prejudice, it tends to be experienced as truth about the world based on scientific findings. Extreme interpretation of Newtonian classical physics is that there is just matter and also the opposite may be true from the extreme interpretations of observing process in the relativity theory and quantum physics in which the observers represent essential entities in the universe.

Both paradigms, classical Newtonian physics as well as modern physics may be connected through the correspondence principle (Jammer 1989; Sells and Weidner 1980), which means that higher level of knowledge includes the previous theory as a special case of this novel theory. The correspondence principle is for example very well described in the relationship between classical Newtonian mechanics and special relativity theory (Einstein 2005; Beiser 1995). Using just very simple mathematical procedures it is possible to obtain basic formulas of classical physics from the special relativity theory. But the opposite is not possible, classical Newtonian mechanics does not include relativity theory as its special case. In this sense a new theory which includes the previous one as its special case represents higher level of knowledge which is not possible to explain as a consequence of logical thinking. In a certain sense this new knowledge is transcendental as a new and different dimension. For example when someone would live on surface, from this point of view (on surface) would not be possible to understand three-dimensional space. In this context, creativity in science is related to discovery of unknown and generating novel ideas. These novel ideas most likely cannot be found on the previous level of interpretation of scientific data just using logical procedures, which is in agreement with Goedel's theorem in mathematics. The Goedel's theorem means that every axiomatic system has basic limits for its logical arguments. This means that for example two logically valid arguments based on axioms and correctly used rules may produce opposite incongruent and paradoxical statements with respect to the same thing (for example this statement is true and this statement is untrue at the same time). Both these statements are valid with respect to the principles of mathematics and logic of the theory but are incongruent and paradoxical. These findings of basic logical principles but also empirical scientific theories show that discoveries and novel ideas transcend logical thinking and represent something that was not before (Penrose 1994; Chaitin 1982).

According to this view scientific invention similarly as artistic work is creative activity that transcends usual viewpoints and represents different and more general view of reality. Creating of this new knowledge is not an observation independent process that produces objective knowledge but in principle it is related to interpretative brain activity based on general principles of attentional selection that tend to connect information contents which are not paradoxical and conflicting, and enable to find congruent cognitive scheme of reality. Anomalies in this picture of reality that lead to paradoxical and unresolved statement later lead to a need to find a new point of view and paradigm (Kuhn 1970). In this context, conflict monitoring and the resolution of the cognitive conflict are based on higher levels of integrative brain activities that within a hierarchically novel contextual schemes enables to resolve paradoxes of the previous cognitive scheme and integrate contents of mind (Bob 2011).

In this sense contextual processing may determine what details of the processed information will actually be presented in conscious awareness while other detailed information related to this percept potentially may be processed and included in other contextual frameworks. Consequently, these contextual mechanisms of attentional sensitivity may influence processing of various aspects of perceptual information and enable to discover events that in certain contextual frameworks are presented as hidden or subliminal parts of the human conscious experience (Baars 2002; Bob 2011). In this sense scientific theory is also the cognitive scheme that enables to find congruent picture of reality.

5.6 Conscious Intuition and Scientific Discoveries

In his basic view of the theory of instinctual behavior Freud (1932/1933) wrote: "The theory of the instincts is so to say our mythology. Instincts are mythical entities, magnificent in their indefiniteness. In our work we cannot for a moment disregard them, yet we are never sure that we are seeing them clearly." The question is how many we can see there and how deep is the human mind in its ontogenetic and fylogenetic history? According to Lorenz (1965) and later findings in ethology this fylogenetic "memory" or contents of information in the unconscious "instinctual" mind are extraordinarily deep. Current behavioral and neuropsychological evidence indicates that these "instinctual" ontogenetically and phylogenetically shared abilities include various levels of adaptive information including highest forms of abstract thinking (Nieder 2005; Dehaene 2009; Cantlon et al. 2009).

In fact we do not know how much information from the fylogenetic history is contained in the human mind. Interesting experiment that uncovers information content of the mind is Ganzfeld phenomenon. Ganzfeld represents mild form of sensory deprivation (Wackerman et al. 2002, 2008) and in principle it is a similar phenomenon like projection tests (for example Rorschach or Thematic Apperception Test) in which ambiguous information contents have various interpretations which enable to release information contained in the human mind and related to previous experiences. Ganzfeld may be easily induced using headphones in which is played white or pink noise (sound similar to noise in radio) and two halves of ping pong ball are used to cover the eyes. At about after 20 min majority of people experience intensive imagination and mild forms of hallucinations in various forms of color scenes and sounds similar to waking dreams (Wackerman et al. 2002, 2008). These data suggest that when the mind is closed with respect to outside stimuli it tends to generate its unconscious and subthreshold contents into conscious awareness similarly as in meditation or during dreams (Hölzel et al. 2007; Ives-Deliperi et al. 2011; Bob 2004).

In this context recent research shows that unconscious mind does not include just primitive forms of instinctual behavior but also highly abstract forms like mathematics. This fylogenetically encoded basis for numerical intuitions and the ability to think in numbers preverbally is evident in human infants, primates and other animal species but the abilities to create abstract numerical concepts are present only in human adults (Feigenson et al. 2004). Overall recent research shows that mathematical intuition is an experimentally testable phenomenon related to extensive cerebral networks (Nieder 2005; Dehaene 2009; Cantlon et al. 2009).

Similarly as in Ganzfeld experiment and in dreams when human mind is closed with respect to outside stimuli the similar kind of "information generation" may be observed in persons who have unusual mathematical intuition. Remarkable cases of "mathematical intuition" were reported in autistic individuals or in individuals with autistic spectrum disorders, who have the so-called "savant syndrome". These cases document outstanding mental capabilities of numerical processing and mathematical intuition coexisting with intellectual limitations and mental disabilities, mainly deficits in social cognition, executive functions, self-involvement and repetitive compulsive behavior (Treffert 1989, 1999, 2002, 2009). An interesting example of the special skills documented Rife and Snyder in 1931, who reported a case of man who was able to calculate in mind the square root of four digit numbers within 4 s and cube root of six digit numbers within 6 s (Rife and Snyder 1931; Hill 1978). Similarly, Heavey et al. (1999) investigated eight savants with calendrical calculation skills and found that the savants did not differ from controls on measures of general short- and long-term memory, nevertheless they had clear superiority for the long-term retention of calendrical material. Interesting epidemiological findings published Rimland (1978), who shown that these special skills in autistic children are not rare. Rimland using detailed questionnaires filled out by the parents of these children studied the incidence of savant skills in 5400 children and found that in 9.8 % were reported some special skills with sex ratio 3.54/1 [males/females].

In many studies have been reported that these autistic savants manifest exceptional abilities for arithmetic or calendrical calculations (Horwitz et al. 1965, 1969; O'Connor and Hermelin 1984; Hermelin and O'Connor 1986, 1990b, 1992; Howe and Smith 1988; Hurst and Mulhall 1988; Ho et al. 1991; Mannheim et al. 1992; Rumsey et al. 1992; Moriarty et al. 1993; Nelson and Pribor 1993; Spitz 1995; Gonzalez-Garrido et al. 2002), musical skills (Viscott 1970; Hermelin et al. 1987, 1989; Miller 1987, 1989, 1995) and other special skills (Morishima and Brown 1977; Brink 1980; Abhyankar et al. 1981; Hermelin and O'Connor 1983, 1990a; O'Connor and Hermelin 1987a, b, 1989, 1990, 1991a, b; Howe 1989a; White 1988; Foerstl 1989; Patti and Lupinetti 1993).

Among these special skills that are closely related to mathematical intuition are particularly interesting abilities of some savants to identify prime numbers that according to mathematical theory have exceptional and central role in arithmetic theory of numbers (Miller 1976; Bressoud 1989). The prime numbers (or primes) are natural numbers (whole and positive) such as 2, 3, 5, 7, 11,..., 83, 89, 97.... that can be divided only by one or itself without a remainder. To find them using mental calculation is a very difficult task mainly for the prime numbers with three or more digits. Interesting case study of savants with especially intriguing ability to identify prime numbers reported Sacks (1985) in 26 years old autistic twins John and Michael in 1966. The twins during their game spontaneously exchanged prime numbers although they did not know simple arithmetic. This task is very difficult also for people with exceptional mathematical skills. A simple method how to identify prime numbers provides Eratosthenes method that enables to find numbers between 2 and n. When n can be for example 9 then all the numbers in the interval are 1, 2, 3, 4, 5, 6, 7, 8, 9 and to find prime numbers needs to delete all numbers that are divisible by two which are 2, 4, 6, 8 and in the second step delete also all numbers divisible by three which are 6 and 9 and what remains after the procedure are the prime numbers 2, 3, 5, 7. These methods work for the prime numbers with 8 digits or less but there is no simple method for calculation of primes of higher orders (Bressoud 1989). But the twins reported by Sacks (1985) during their game with primes at first created 6-digit primes that they created within seconds and then also larger 10-digit prime numbers that they created within 5 min and later during the game they exceeded tables of prime numbers that were available during this "experiment". These extraordinary abilities were confirmed also in other studies and also with savants who had not any arithmetical skills that would be needed to perform calculations (Anderson et al. 1999; Welling 1994).

According to some recent findings these extraordinarily skills might be linked to abnormalities in activities of the right cerebral hemisphere that may be specifically influenced by insufficient activity of the left hemisphere (Delong 1999; Gazzaniga 2000; Corballis et al. 2000). In this context, interesting hypothesis about neurophysiological nature of the savant skills suggested Snyder (2009) based on research data that savant skills may be artificially induced by partial inactivation of the left hemisphere using low-frequency repetitive transcranial magnetic stimulation (rTMS) in healthy people that may temporarily inhibit neural activity in some cortical areas and create "virtual lesions" (Hilgetag et al. 1999; Walsh and Cowey 2000; Hoffman and Cavus 2002) and decrease top-down inhibitory influences which may enable to have access to lower levels of information processes as savants do (Snyder 2009).

Snyder (2009) proposed that savant-like skills may be also artificially induced in normal healthy individuals by inhibiting the left anterior temporal lobe in agreement with data that autistic savants have left brain dysfunction and atypical inhibitory functions of the left hemisphere (Wilson et al. 2007) that are compensated by the right hemisphere (Miller et al. 1998; Treffert 2005; Sacks 2007), which leads to a predilection for literal, non-symbolic skills (Sacks 2007, pp. 314–315; Treffert 2005, 2006). This explanation is also in agreement with the right-hemispheric bias that is frequently associated with autism (Herbert et al. 2005; Koshino et al. 2005). In support of this hypothesis Snyder et al. (2003) have found significant changes in drawing style following active rTMS stimulation. The changes in drawing were characterized by heightened focus on seemingly meaningless details and naturalistic drawing style which may be explained by heightened access to raw, less-processed sensory information typical for savants' performance (Snyder and Mitchell 1999; Snyder 2009), which may be caused by sensory hypersensitivity and enhanced perception of details (Minshew and Hobson 2008). This increased sensitivity and perception likely exists not only in savants but also in normal individuals who may have the increased access to this information because of decreased top-down inhibition (Snyder et al. 2003). Similar data reported also Young et al. (2004) Miller et al. (1996, 1998) who found increased awareness of detail which are typical for savant-like skills.

This sensitivity to detail related to the inhibitory failure likely may explain unusual ability of some autistic savants to process mathematical information in a way indicating mathematical intuition but without normal ability to integrate these spontaneous insights into concepts that enable to integrate fragmentary data in mutual connections. This mode of mental processing likely reflects also specific changes in brain information processing which according to recent evidence is less globally integrated with long-distance under-connectivity and with higher level of integration on some local levels leading to local over-connectivity (Wass 2011; Nomura et al. 2010; Parmeggiani et al. 2010; Tuchman et al. 2010).

This process of focus on details likely may enable unusual intuition and imaginative capacity. In this sense, Weilling suggested that a distinction between prime and nonprime numbers could be made by an ability to recognize the primes through a spontaneous tendency to find symmetry (Welling 1994). Intriguing example how this tendency and ability to find symmetry may be performed shows autistic savant Daniel Tammet, who is well-known by his European record for the number of digits of pi that he was able to know from memory [i.e. 22,514 digits] (Biever 2009). But in contrast to many other savants he was able to describe how he does it. Daniel Tammet reported his subjective experiences to see numbers in spatial forms with sizes, colours and texture where prime numbers have special properties (like color, position etc.) that distinguish them from other numbers (Tammet 2006; Baron-Cohen et al. 2007; Murray 2010). Based on similar data Azouli et al. (2005) proposed synesthesia as a possible mechanism of savant skills.

Together these data suggest that the ability to recognize prime numbers may present intuititive intelligence reflecting mathematically exceptional and fundamental role of the prime numbers and their sequences in mathematics and physics (for example see Miller 1976; Agrawal et al. 2004; Iovane et al. 2004; Iovane 2008). Although some autistic savants demonstrate unusual examples of mathematical intuition and likely have insights into mathematical symmetry, typical deficit of these individuals in their processing of mathematical information is related to inability to integrate fragmentary insights into mutual connections and create abstract numerical and mathematical concepts.

Intriguing ability of these savants to process details is likely based on changes in brain mechanisms that enable multilevel information processing regulated within the framework of contextual understanding which strictly determines what details are important for the whole coherent context (Mesulam 1998; Nadel and Jacobs 1998; Lavenex and Amaral 2000; Hupe et al. 1998; Bullier 2006; Fabricius 2010). These findings suggest that the multilevel information contextual processing of perceptual information might be related to executive signals that in the framework of required context determine sensitivity to various details needed for contextual processing (Gilbert and Wiesel 1990; Francis et al. 1994; Gove et al. 1995; Gray 1999; Grossberg and Grunewald 1997; Ito and Gilbert 1999; Raizada and Grossberg 2001; Kang et al. 2005).

On the other hand, in the similar situation is also mind of mathematicians and scientists, who are also extremely focused, contemplative and autistic-like which might explain why some great mathematicians and scientists like Newton, Einstein and some others were thought to have mild form of autism or Asperger syndrome (Asperger 1991; James 2003; Hazel 2003; Fitzgerald 2004). But independently of what we can know about the intuitive nature of mathematical knowledge, the interaction between mind and brain enables to discover laws of nature and understand the external world through the rules that connect observed objects and processes.

Typical example provides Einstein (1918), who expressed his personal view: "The supreme task of the physicist is to arrive at those universal elementary laws from which the cosmos can be built up by pure deduction. There is no logical path to these laws; only intuition, resting on sympathetic understanding of experience, can reach them."

The role of intuition in scientific knowledge is particularly interesting in some cases of important mathematical theories that were established without direct connection to experimental science and their discovery was based mainly on mathematical intuition and abstraction. Exceptional examples of such mathematical theories in the history of science are differential and integral calculus representing mathematical basis for Newtonian formulation of mechanics and few centuries later developed Riemanian geometry. For example, Riemannian geometry (Plotnitsky 2009) was discovered at about 50 years before the Einstein's discovery of general relativity theory (Laue 1950). And without general relativity would not be possible to know anything about the black holes and other objects that are significantly different than usually expected behavior described within the framework of Newtonian mechanics.

These modern findings in cosmology show incomparably different view of the universe than Newton and Descartes proposed. In comparison to Newtonian mechanical universe including bodies with known basic physical characteristics, recent cosmology has found that just only smaller amount of matter in the universe is the "usual" matter that constitutes objects like stars, planets and other space objects that have observable physical characteristics and interact with light and other frequencies of the electromagnetic field. According to these modern findings mainly based on general relativity theory significant majority of matter in the universe appears to be the so-called dark matter which has only mass and gravitation but likely does not have other known physical characteristics of matter and does not interact with electromagnetic field and [similarly like black holes] cannot be directly observed (Carmeli 1999; Capozziello and Sarkar 2005; Capozziello and Francaviglia 2008; Zakharov et al. 2009; Bertone 2010).

The difference between Newtonian concept of the universe and relativity theory does not mean just an improvement of the old theoretical concept but deep change in principle, which means that using mathematics and logical thinking it is not possible to obtain relativity theory from Newtonian mechanics, but on the other hand it is very easy to deduce Newtonian mechanics from the special relativity theory (Laue 1950; Pauli 1958; Einstein 2005).

This mystery of discovery represents also Freud's "Project" applying basic thermodynamic principles on brain and mind with the aim to find connections of the physical theory with the processes studied in psychology. According to current knowledge laws of thermodynamics represent fundamental rules applicable in physics, chemistry, biology, economy and other interdisciplinary fields of research (Nicolis and Prigogine 1977; Sieniutycz and Salamon 1990; Epstein and Pojman 1998; Perrot 1998; Erdi 2007). From 1980s thermodynamics and its relatively newly developed fields of research related to nonlinear thermodynamics such as complexity theory, chaos theory and theory of self-organization have been also applied in brain research and psychology (Elbert et al. 1994; Freeman 1983, 1991, 2000; Barton 1994; Birbaumer et al. 1995; Meyer-Lindenberg et al. 2002; Faure and Korn 2001; Globus and Arpaia 1994; Korn and Faure 2003).

As a pioneer in this research, Freud followed Brücke's (1873–1874) general concept that all living organisms are in fact energy-systems similar to non-living systems with respect to thermodynamic laws. In this context, Freud applied basic principles of thermodynamics and used the term "psychic energy" and proposed that in principle this concept may be applied also for description of the mind. This "Project" that Freud (1895) in his work outlined is in accordance with current concepts in physics of entropy and information theory that together with neuroscientific research present new promising perspectives for psychological science. In this context, basic functioning of the mind might be understood on 'thermodynamic' principles of energy exchange, which similarly like all physical systems, for example atoms that have free energy to move or create molecules, the mind has free energy to create spontaneous intentional association patterns like fantasies which generate typical patterns of behavior and relationships.

In this context the Freud's "Project" although unfinished is a remarkable work focused on application of general thermodynamic principles in the field of psychology and neuroscience. This work is mainly intuitive view that within neural-thermodynamic framework anticipated certain findings of recent nonlinear thermodynamic theory of the brain functions.

Certain data from his "Project" also suggest that Freud mentioned several general concepts that was later used for theoretical explanations in memory research linked to the term "long term potentiation" when he described phenomenon of summation of neuronal activities and representing memory in the brain as permanent alterations following an event (Centonze et al. 2004). In similar context McClelland (1998) described relationships between basic Freud's concept proposed in the "Project" and connectionist approach to explicit and implicit cognition and memory. From the contemporary point of view thermodynamic principles used in Freud's "Project" seem to be relevant and represent conceptually perspective view for the future research and intuitively anticipate several current findings about neural systems and their specific interactions.

Most enigmatic and mysterious element of the Freud's psychic apparatus are "omega neurons" producing attentional functions, self-reference and conscious experience per se. Stephen Hawking in his "Short History of Time" wrote that the real theory of everything needs also to explain its own origin, which means to describe the process of its own creation in the human mind. Does it mean that all this work may be done by omega neurons, and we are omega neurons which through the human mind recognize themselves in the microscope. Or are we vehicles of genes that manipulate with us and use the human mind to recognize themselves,…? In the sense of Freud's theory, it is the transference which causes that we project our own mental states onto other things in the Universe and create mythological, religious and scientific paradigms that reflect unconscious aspects of our mind.

Does this process of sublimation and self-recognition mean that ability to discover our-selves is possible through the view of the abstract mind as a basis for self-reference and integrative mental experience which is the same in all observers? And when consciousness as an abstract mind creates background of unity, is it possible that it can be 'divided' into plurality of minds or it is the same mind in everyone.

…

Sigmund Freud never finished his "Project" and this work meant for him a lot of personal troubles. In the time when he worked on this project he dreamt early in the morning of July 24, 1895 this dream ("Dream of Irma's Injection", Freud 1991, p. 182):

> A large hall—numerous guests, whom we were receiving.—Among them was Irma. I at once took her on one side, as though to answer her letter and to reproach her for not having accepted my 'solution' yet. I said to her: 'If you still get pains, it's really only your fault.' She replied: 'If you only knew what pains I've got now in my throat and stomach and abdomen—it's choking me.'—I was alarmed and looked at her. She looked pale and puffy. I thought to myself that after all I must be missing some organic trouble. I took her to the

> window and looked down her throat, and she showed signs of recalcitrance, like women with artificial dentures. I thought to myself that there was really no need for her to do that.—She then opened her mouth properly and on the right hand I found a big white patch; at another place I saw extensive whitish grey scabs upon some remarkable curly structures which were evidently modeled on the turbinal bones of the nose.—I at once called in Dr. M., and he repeated the examination and confirmed it ..... Dr. M. looked quite different from usual; he was very pale, he walked with a limp and his chin was clean-shaven ..... My friend Otto was now standing beside her as well, and my friend Leopold was percussing her through her bodice and saying: 'She has a dull area low down on the left.' He also indicated that a portion of the skin on the left shoulder was infiltrated. (I noticed this, just as he did, in spite of her dress.) .... M. said: 'There's no doubt it's an infection, but no matter; dysentery will supervene and the toxin will be eliminated.' ... We were directly aware, too, of the origin of the infection. Not long before, when she was feeling unwell, my friend Otto had given her an injection of a preparation of propyl, propyls ... propionic acid ... trimethylamin (and I saw before me the formula for this printed in heavy type) .... Injections of that sort ought not to be made so thoughtlessly .... And probably the syringe had not been clean.

In the morning July 24, 1895 after this dream Sigmund Freud wrote a letter to his friend Wilhelm Flies (Mason 1895): "Daimonie [demon], why don't you write? How are you? Don't you care at all any more about what I am doing? What is happening to the nose, menstruation, labor pains, neuroses, your dear wife, and the budding little one? True, this year I am ill and must come to you; what will happen if by chance both of us remain healthy for a whole year? are we friends only in misfortune? Or do we also want to share the experiences of calm with each other. Where will you spend the month of August? We are living very contentedly in Himmel."

References

Abhyankar, R., Thatte, S., & Doongaji, D. (1981). Idiot savant. *Journal of Postgraduate Medicine, 22*, 44–47.

Acheson, A., Conover, J. C., Fandl, J. P., DeChiara, T. M., Russell, M., Thadani, A., Slinto, S. P., Yancopoulos, G. D., & Lindsay, R.M. (1995). A BDNF autocrine loop in adult sensory neurons prevents cell death. *Nature, 374*, 450–3.

Adamec, R. (1997). Transmitter systems involved in neural plasticity underlying increased anxiety and defense-implications for understanding anxiety following traumatic stress. *Neuroscience and Biobehavioral Reviews, 21*, 755–765.

Adamec, R. E. (1990). Does kindling model anything clinically relevant? *Biological Psychiatry, 27*, 249–279.

Adamec, R. E. (1999). Evidence that limbic neural plasticity in the right hemisphere mediates partial kindling induced lasting increases in anxiety-like behavior: Effects of low frequency stimulation (Quenching?) on long-term potentiation of amygdala efferents and behavior following kindling. *Brain Research, 839*, 133–152.

Adamec, R. E., & McKay, D. (1993). Amygdala kindling, anxiety, and corticotopin releasing factor (CRF). *Physiology and Behavior, 54*, 423–431.

Aftanas, L. I., & Golocheikine, S. A. (2002). Non-linear dynamic complexity of the human EEG during meditation. *Neuroscience Letters, 330*, 143–146.

Agargun, M. Y., Tekeoglu, I., Kara, H., Adak, B., & Ercan, M. (1998). Hypnotizability, pain threshold, and dissociative experiences. *Biological Psychiatry, 44*, 66–71.

Agrawal, M., Kayal, N., & Saxena, N. (2004). PRIMES is in P. *Annals of Mathematics, 160*, 781–793.

Agre, P. E. (1997). *Computation and human experience*. Cambridge: Cambridge University Press.

Ahern, G. L., Herring, A. M., Tackenberg, J., & Seeger, J. F. (1993). The association of multiple personality and temporolimbic epilepsy: Intracarotid amobarbital test observations. *Archives of Neurology, 50*, 1020–1025.

Aks, D. J., & Sprott, J. C. (2003). Resolving perceptual ambiguity in the Necker Cube: A dynamical systems approach. *Journal of Non-linear Dynamics in Psychology and the Life Sciences, 7*, 159–178.

Alarcon, G., Binnie, C. D., Elwes, R. D., & Polkey, C. E. (1995). Power spectrum and intracranial EEGpatterns at seizureonset in partial epilepsy. *Electroencephalography and Clinical Neurophysiology, 94*, 326–337.

Albert, D. (1992). *Quantum mechanics and experience*. Cambridge: Harvard University Press.

Albright, P. S., & Burnham, B. M. (1980). Development of a new pharmacological seizure model: Effects of anticonvulsants on cortical and amygdala-kindled seizures in the rats. *Epilepsia, 21*, 681–689.

P. Bob, *The Brain and Conscious Unity*, DOI 10.1007/978-1-4939-2700-5

Alison, R. B. (1974). A new treatment approach for multiple personalities. *American Journal of Clinical Hypnosis, 17*, 15–32.

Alkire, M. T. (2008). Loss of effective connectivity during general anesthesia. *International Anesthesiology Clinics, 46*, 55–73.

Allman, J. M., Hakeem, A., Erwin, J. M., Nimchinsky, E., & Hof, P. (2001). The anterior cingulate cortex. The evolution of an interface between emotion and cognition. *Annals of the New York Academy of Sciences, 935*, 107–117.

American Psychiatric Association. (1987). *DSM III-R, diagnostic and statistical manual of mental disorders* (4th ed.). Washington, DC: American Psychiatric Association.

American Psychiatric Association. (1994). *DSM IV, diagnostic and statistical manual of mental disorders* (4th ed.). Washington, DC: American Psychiatric Association.

Anderson, C. M., Teicher, M. H., Polcari, A., & Renshaw, P. F. (2002). Abnormal T2 relaxation time in the cerebellar vermis of adults sexually abused in childhood: Potential role of the vermis in stress-enhanced risk for drug abuse. *Psychoneuroendocrinology, 27*, 231–244.

Anderson, J. R. (1995). *Cognitive psychology and its implications*. New York: Freeman.

Anderson, M., O'Connor, N., & Hermelin, B. (1999). A specific calculating ability. *Intelligence, 26*, 383–403.

Andersson, O. (1962). *Studies in the prehistory of psychoanalysis*. Stockholm: Svenska Bokforlaget.

Andreasen, N. C. (1997). Linking mind and brain in the study of mental illnesses: A project for a scientific psychopathology. *Science, 275*, 1586–1593.

Arp, R. (2005). Selectivity, integration, and the psycho-neuro-biological continuum. *Journal of Mind and Behavior, 26*, 35–64.

Ashby, W. R. (1947). Principles of the self-organizing dynamic system. *Journal of General Psychology, 37*, 125–128.

Aspect, A. (1999). Bell's inequality test: More ideal than ever. *Nature, 398*, 189–190.

Aspect, A., Dalibard, J., & Roger, G. (1982). Experimental test of Bell's inequalities using time-varying analyzers. *Physical Review Letters, 49*, 1804–1807.

Asperger, H. (1991). Autistic psychopathy in childhood (U. Frith, Trans.). In U. Frith (Ed.), *Autism and Asperger syndrome* (pp. 37–92). Cambridge: Cambridge University Press (Original work published 1944).

Atmanspacher, H., & Fach, W. (2005). Acategoriality as mental instability. *Journal of Mind and Behavior, 26*, 181–205.

Avila, E. K., & Graber, J. (2010). Seizures and epilepsy in cancer patients. *Current Neurology and Neuroscience Report, 10*, 60–67.

Azouli, S., Hubbard, E. M., & Ramachandran, V. S. (2005). Does synesthesia contribute to mathematical savant skills? *Journal of Cognitive Neuroscience, 69*, 1 (Supl). http://psy2.ucsd.edu/~edhubbard/abstracts/Azoulai_CNS05.pdf.

Baars, B. J. (1988). *A cognitive theory of consciousness*. Cambridge: Cambridge University Press.

Baars, B. J. (1997). Some essential differences between consciousness and attention, perception, and working memory. *Consciousness and Cognition, 6*, 363–371.

Baars, B. J. (1999). Attention versus consciousness in the visual brain: Differences in conception, phenomenology, behaviour, neuroanatomy, and physiology. *Journal of General Psychology, 12*, 224–233.

Baars, B. J. (2002). The conscious access hypothesis: Origins and recent evidence. *Trends in Cognitive Sciences, 6*, 47–52.

Balduzzi, D., & Tononi, G. (2008). Integrated information in discrete dynamical systems: Motivation and theoretical framework. *PLoS Computational Biology, 4*(6), e1000091.

Balian, R. (2003). Entropy—Protean concept. *Poincaré Seminar, 2*, 119–145.

Ballentine, L. E., & Jarrett, J. (1997). Bell's theorem: Does quantum mechanics contradict relativity? *American Journal of Physics, 55*, 785–792.

Bancaud, J., Brunet-Bourgin, F., Chauvel, P., & Halgren, E. (1994). Anatomical origin of déjà vu and vivid 'memories' in human temporal lobe epilepsy. *Brain, 117*, 71–90

Bao, A. M., Meynen, G., & Swaab, D. F. (2008). The stress system in depression and neurodegeneration: Focus on the human hypothalamus. *Brain Research Reviews, 57*, 531–553.

Baron-Cohen, S., Bor, D., Billington, J., Asher, J., Wheelwright, S., & Ashwin, C. (2007). Savant memory in a man with colour form-number synaesthesia and asperger syndrome. *Journal of Consciousness Studies, 14*, 237–251.

Barrera-Mera, B., & Barrera-Calva, E. (1998). The Cartesian clock metaphor for pineal gland operation pervades the origin of modern chronobiology. *Neuroscience and Biobehavioral Reviews, 23*, 1–4.

Barret, D. (1994). Dreams in dissociative disorders. *Dreaming, 4*, 165–175.

Barret, D. (1995). The dream character as prototype for the multiple personality alter. *Dissociation, 8*, 66–68.

Barret, D. (1996). Dreams in multiple personality disorder. In D. Barret (Ed.), *Trauma and dreams* (pp. 68–81). Cambridge: Harvard University Press.

Barrow-Green, J. (1997). *Poincaré and the three body problem*. Providence: American Mathematical Society.

Bartels, A., & Zeki, S. (2006). The temporal order of binding visual attributes. *Vision Research, 46*, 2280–2286.

Barton, S. (1994). Chaos, self-organization, and psychology. *American Psychologist, 49*, 5–14.

Bass, L. (1975). A quantum-mechanical mind-body interaction. *Foundations of Physics, 5*, 159–172.

Bath, K. G., & Lee, F. S. (2006). Variant BDNF (Val66Met) impact on brain structure and function. *Cognitive Affective and Behavioral Neuroscience, 6*, 79–85.

Bauer, J., & Bien, C. G. (2009). Encephalitis and epilepsy. *Seminars in Immunopathology, 31*, 537–544.

Baydas, G., Ozveren, F., Akdemir, I., Tuzcu, M., & Yasar, A. (2005). Learning and memory deficits in rats induced by chronic thinner exposure are reversed by melatonin. *Journal of Pineal Research, 39*, 50–56.

Bechara A., Tranel, D., Damasio, H., Adolphs, R., Rockland, C., & Damasio, A. R. (1995). Double dissociation of conditioning and declarative knowledge relative to the amygdala and hippocampus in humans. *Science, 269*, 1115–1118.

Beck, F., & Eccles, J. C. (1992). Quantum aspects of brain activity and the role of consciousness. *Proceedings of the National Academy of Sciences of the United States of America, 89*, 11357–11361.

Beiser, A. (1995). *Concepts of modern physics* (5th ed.). New York: Mc Graw Hill.

Bell, J. (1987). *Speakable and unspeakable in quantum mechanics*. Cambridge: Cambridge University Press.

Benarroch, E. E. (1993). The central autonomic network—Functional organization, dysfunction and perspective. *Mayo Clinic Proceedings, 68*, 988–1001.

Benes, F. M., & Berretta, S. (2001). GABAergic interneurons: Implications for understanding schizophrenia and bipolar disorder. *Neuropsychopharmacology, 25*, 1–27.

Benson, D. F. (1986). Interictal behaviour disorders in epilepsy. *Psychiatric Clinics of North America, 9*, 283–292.

Benson, F., Miller, B. L., & Signer, S. F. (1986). Dual personality associated with epilepsy. *Archives of Neurology, 43*, 471–474.

Bernfeld, S. (1949). Freud's scientific beginnings. *American Imago, 6*, 163–196.

Bernstein, E. M., & Putnam, F. W. (1986). Development, reliability, and validity of a dissociation scale. *Journal of Nervous and Mental Disease, 174*, 727–735.

Bertone, G. (2010). The moment of truth for WIMP dark matter. *Nature, 468*, 389–393.

Berut, A., Arakelyan, A., Petrosyan, A., Ciliberto, S., Dillenschneider, R., & Lutz, E. (2012). Experimental verification of Landauer's principle linking information and thermodynamics. *Nature, 483*, 187–189.

Besedovsky, H., Del Rey, A., Sorkin, E., & Dinarello, C. A. (1986). Immunoregulatory feedback between interleukin-1 and glucocorticoid hormones. *Science, 233*, 652–654.

Beutel, M. E., Dietrich, S., Stark, R., Brendel, G., & Silbersweig, D. (2004). Pursuit of the emerging dialogue between psychoanalysis and neuroscience: Clinical and research perspectives. *International Journal of Psychoanalysis, 85*, 1493–1496.

Biever, C. (2009). Peek inside a singular mind. New Scientist, 2689, pp. 40–41. Online version: Inside the mind of an autistic savant. New Scientist. http://www.newscientist.com/article/mg20126881.800-inside-the-mind-of-an-autistic-savant.html.

Binder, D. K., & Scharfman, H.E. (2004). Brain-derived neurotrophic factor. *Growth Factors, 22*, 123–131.

Birbaumer, N., Flor, H., Lutzenberger, W., & Elbert, T. (1995). Chaos and order in the human brain. *Electroencephalography and Clinical Neurophysiolology* (Suppl.), *44*, 450–459.

Birbaumer, N., Lutzenberger, W., & Elbert, T. (1993). Imagery and brain processes. In N. Birbaumer & A. Ohman (Eds.), *The structure of emotion* (pp. 122–138). Toronto. Hogrefe and Huber.

Bizas, E., Simos, P. G., Stam, C. J., Arvanitis, S., Terzakis, D., & Micheloyannis, S. (1999). EEG correlates of cerebral engagement in reading tasks. *Brain Topography, 12*, 99–105.

Blake, R., Yu, K., Lokey, M., & Norman, H. (1998). Binocular rivalry and motion perception. *Journal of Cognitive Neuroscience, 10*, 46–60.

Blanke, O., Ortigue, S., Landis, T., & Seeck, M. (2002). Neuropsychology: Stimulating illusory own-body perceptions. *Nature, 419*, 269–270.

Bleuler, E. (1911/1955). *Dementia praecox or the group of the schizophrenias*. New York: International University Press.

Bleuler, E. (1918/1906). Consciousness and association (M. D. Eder (trans.)). In C. G. Jung (Ed.), *Studies in word-association* (pp. 266–296). London: William Heinemann.

Bleuler, E. (1924). *Textbook of psychiatry* (A. A Brill (trans.)). New York: Macmillan Publishing Co. Inc.

Bliss, E. L. (1980). Multiple personalities. A report of 14 cases with implications for schizophrenia and hysteria. *Archives of General Psychiatry, 37*, 1388–1397.

Blumer, D., Wakhlu, S., Davies, K., & Hermann, B. (1998). Psychiatric outcome of temporal lobectomy for epilepsy: Incidence and treatment of psychiatric complications. *Epilepsia, 39*, 478–486.

Boatright, J. H., Rubim, N. M., & Iuvone, P. M. (1994). Regulation of endogenous dopamine release in amphibian retina by melatonin: The role of GABA. *Visual Neuroscience, 11*, 1013–1018.

Bob, P. (2003). Dissociation and neuroscience: History and new perspectives. *International Journal of Neuroscience, 113*, 903–914.

Bob, P. (2003a). Subliminal processes dissociation and the 'I'. *Journal of Analytical Psychology, 48*, 307–316.

Bob, P. (2004). Dissociative processes, multiple personality and dreamfunctions. *American Journal of Psychotherapy, 58*, 139–149.

Bob, P. (2007). Hypnotic abreaction releases chaotic patterns of electrodermal activity during dissociation. *International Journal of Clinical and Experimental Hypnosis, 55*, 435–456.

Bob, P. (2008). Pain, dissociation and subliminal self-representations. *Consciousness and Cognition, 17*, 355–369.

Bob, P. (2009). Quantum science and the nature of mind. *Journal of Mind and Behavior, 30*, 1–14.

Bob, P. (2011). *Brain, mind and consciousness: Advances in neuroscience research*. New York: Springer.

Bob, P. (2012). Consciousness, schizophrenia and complexity. *Cognitive Systems Research, 13*(1), 87–94.

Bob, P., & Fedor-Freybergh, P. (2008). Melatonin, consciousness and traumatic stress. *Journal of Pineal Research, 44*, 341–347.

Bob, P., & Svetlak, M. (2011). Dissociative states and neural complexity. *Brain and Cognition, 75*(2), 188–195.

Bob, P., Susta, M., Gregusova, A., & Jasova, D. (2009a). Dissociation, cognitive conflict and nonlinear patterns of heart rate dynamics in patients with unipolar depression. *Progress in Neuropsychopharmacology and Biological Psychiatry, 33*, 141–145.

Bob, P., Susta, M., Chladek, J., Glaslova, K., Palus, M. (2009b). Chaos in schizophrenia associations, reality or metaphor? *International Journal of Psychophysiology, 73*, 179–185.

Bob, P., Susta, M., Glaslova, K., & Boutros, N. N. (2010). Dissociative symptoms and interregional EEG cross-correlations in paranoid schizophrenia. *Psychiatry Research, 177*, 37–40.

Bob, P., Palus, M., Susta, M., Glaslova, K. (2010a). Sensitization, epileptic-like symptoms and local synchronization in patients with paranoid schizophrenia. *Progress in Neu-Psychopharmacology and Biological Psychiatry, 34*, 143–146.

Bob, P., Susta, M., Gregusova, A., Jasova, D., Mishara, A., & Raboch, J. (2010b). Traumatic stress, dissociation, and limbic irritability in patients with unipolar depression being treated with SSRIs. *Psychological Reports, 107*, 685–696.

Bob, P., Raboch, J., Maes, M., Susta, M., Pavlat, J., Jasova, D., Vevera, J., Uhrova, J., Benakova, H., & Zima, T. (2010c). Depression, traumatic stress and interleukin-6. *Journal of Affective Disorders, 120*, 231–234.

Bob, P., Golla, M., Epstein, P., & Konopka, L. (2011). EEG complexity and attentional processes related to dissociative states. *Clinical EEG and Neuroscience, 42*, 175–179.

Bogdan, R., & Pizzagalli, D. A. (2006). Acute stress reduces reward responsiveness: Implications for depression. *Biological Psychiatry, 60*, 1147–1154.

Bogen, J. E. (1995). On the neurophysiology of consciousness: Part I-an overview. *Consciousness and Cognition, 4*, 52–62.

Bogen, J. E., & Bogen, G. M. (1969). The other side of the brain: III. The corpus callosum and creativity. *Bulletin of Los Angeles Neurological Society, 34*, 191–220.

Bohr, N. (1958). *Atomic physics and human knowledge*. New York: Wiley.

Bower, G. H. (1981). Mood and memory. *American Psychologist, 36*, 129–148.

Bowers, K. S., & Brenneman, H. A. (1979). Hypnosis and the perception of time. *International Journal of Clinical and Experimental Hypnosis, 27*, 29–41.

Bowers, M. K., & Brecher, S. (1955). The emergence of multiple personalities in the course of hypnotic investigation. *International Journal of Clinical and Experimental Hypnosis, 3*, 188–199.

Bowman, E. S., & Coons, P. M. (2000). The differential diagnosis of epilepsy, pseudoseizures, dissociative identitydisorder, and dissociative disorder not otherwise specified. *Bulletin of the Menninger Clinic, 64*, 164–180.

Boyer, C. B. (1968). *History of mathematics*. New York: Wiley.

Brackenridge, J. B. (1995). *The key to Newton's dynamics: The Kepler problem and the principia*. Berkeley: University of California Press.

Braitenberg, V. (1978). Cell assemblies in the cerebral cortex. In R. Heim & G. Palm (Eds.), *Theoretical approaches to complex systems. Lecture notes in biomathematics* (pp. 171–188). Berlin: Springer.

Bramham, C. R., & Messaoudi, E. (2005). BDNF function in adult synaptic plasticity: The synaptic consolidation hypothesis. *Progress in Neurobiology, 76*, 99–125.

Brannon, E. M. (2006). The representation of numerical magnitude. *Current Opinion in Neurobiology, 16*, 222–229.

Braun, B. G. (1986). Issues in the psychotherapy of multiple personality disorder. In B. G. Braun (Ed.), *Treatment of multiple personality disorder* (pp. 1–28). Washington, DC: American Psychiatric Press.

Brazdil, M., Rektor, I., Daniel, P., Dufek, M., & Jurak, P. (2001). Intracerebral event-related potentials to subthreshold target stimuli. *Clinical Neurophysiology, 112*, 650–661.

Breakspear, M. (2006). The nonlinear theory of schizophrenia. *Australian and New Zealand Journal of Psychiatry, 40*, 20–35.

Bremner, J. D. (1999). Does stress damage the brain. *Biological Psychiatry, 45*, 797–805.

Bremner, J. D. (2006). Stress and brain atrophy. *CNS and Neurological Disorders—Drug Targets, 5*, 503–512.

Bremner, J. D. (2006). The relationship between cognitive and brain changes in posttraumatic stress disorder. *Annals of the New York Academy of Sciences, 1071*, 80–86.

Bremner, J.D., Randall, P., Scott, T. M., Bronen, R. A., Seibyl, J. P., Southwick, S. M., Delaney, R. C., & Charney, D. S. (1995). MRI-based measurement of hippocampal volume in combat-related posttraumatic stress disorder. *American Journal of Psychiatry, 152*, 973–981.

Bremner, J. D., Elzinga, B., Schmahl, C., & Vermetten, E. (2007). Structural and functional plasticity of the human brain in posttraumatic stress disorder. *Progress in Brain Research, 167*, 171–186.

Brende, J. O. (1984). The psychophysiologic manifestations of dissociation: Electrodermal responses in a multiple personality patient. *Psychiatric Clinics of North America, 7*, 41–50.
Brende, J. O. (1982). Electrodermal responses in posttraumatic syndromes. A pilot study of cerebral hemisphere functioning in Vietnam veterans. *Journal of Nervous and Mental Disease, 170*, 352–361.
Brenner, I. (1996). The characterological basis of multiple personality. *American Journal of Psychotherapy, 52*, 154–166.
Brenner, I. (1999). Deconstructing DID. *American Journal of Psychotherapy, 53*, 344–360.
Brenner, I. (2001). *Dissociation of trauma: Theory, phenomenology and technique* (p. 9). Boston: International Universities Press.
Bressler, S. L. (1995). Large-scale cortical networks and cognition. *Brain Research Reviews, 20*, 288–304.
Bressler, S. L., Coppola, R., & Nakamura, R. (1993). Episodic multiregional cortical coherence at multiple frequencies during visual task performance. *Nature, 366*, 153–156.
Bressoud, D. M. (1989). *Factorizations and primality testing*. New York: Springer.
Breuer, J., & Freud, S. (1895). *Studies in hysteria*. New York: Basic Books.
Brewin, C. R. (2007). Autobiographical memory for trauma: Update on four controversies. *Memory, 15*, 227–248.
Briere, J., & Conte, J. (1989). *Amnesia and adults molested as children: Testing theories of repression*. Paper presented at the 97th Annual Convention of the American Psychological Association, New Orleans.
Brillouin, L. (1956). *Science and information theory*. New York: Academic Press.
Brink, T. (1980). Idiot savant with unusual mechanical ability: An organic explanation. *American Journal of Psychiatry, 137*, 250–251.
Briquet, P. (1859). *Traité clinique et therapeutique de l'hystérie*. Paris: JB Baillière
Brown, B. G. (1984). Towards a theory of multiple personality and other dissociative phenomena. *Psychiatric Clinics of North America, 7*, 171–193.
Brown, D., & Fromm, E. (1986). *Hypnotherapy and hypnoanalysis*. Hillsdale: L. Erlbaum Associates.
Brown, R. J., & Trimble, M. R. (2000). Dissociative psychopathology, non-epileptic seizures, and neurology. *Journal of Neurology, Neurosurgery and Psychiatry, 69*, 285–9.
Brücke, E. W. (1873–1874). *Vorlesungen über Physiologie*. Vienna: Wilhelm Braumüller.
Buchwald, J. (1989). *The rise of the wave theory of light: Optical Theory and experiment in the early nineteenth century*. Chicago: University of Chicago Press.
Bullier, J. (2006). What is fed back? In J. van Hemmen & T. Senjowski (Eds.), *Problems in systems neuroscience* (pp. 103–132). New York: Oxford University Press.
Bunge, S. A., Ochsner, K. N., Reskond, J. E., Dover, G. H., Gabrieli, J. D. E. (2001). Prefrontal regions involved in keeping information in and out of mind. *Brain, 124*, 2074–2086.
Butler, L. D., Duran, R. E. F., Jasiukaitis, P., Koopman, C. H., & Spiegel, D. (1996). Hypnotizability and traumatic experience: A diathesis-stress model of dissociative symptomatology. *American Journal of Psychiatry, Festschrift Supplement, 153*, 42–62.
Buzsaki, G. (2006). *Rhythms of the brain*. Oxford: Oxford University Press.
Cahill, L. (1997). The neurobiology of emotionally influenced memory. Implications for understanding traumatic memory. *Annals of the New York Academy of Sciences, 821*, 238–246.
Cahill, L., & McGaugh, J. L. (1998). Mechanisms of emotional arousal and lasting declarative memory. *Trends in Neurosciences, 21*, 294–299.
Calabrese, F., Molteni, R., Racagni, G., & Riva, M. A. (2009). Neuronal plasticity: A link between stress and mood disorders. *Psychoneuroendocrinology, 34*(Suppl 1), S208–216.
Callender, C., & Huggett, N. (2001). *Physics meets philosophy at the planck scale*. Cambridge: Cambridge University Press.
Camazine, S., Deneubourg, J. L., Franks, N., Sneyd, J., Theraulaz, G., & Bonabeau, E. (2003). *Self-organization in biological systems*. Princeton: Princeton University Press.
Cantlon, J. F., Platt, M. L., & Brannon, E. M. (2009). Beyond the number domain. *Trends in Cognitive Sciences, 13*, 83–91.

Capozziello, S., & Francaviglia, M. (2008). Extended theories of gravity and their cosmological and astrophysical applications. *General Relativity and Gravitation, 40*, 357–420.

Carmeli, M. (1999). Aspects of cosmological relativity. *International Journal of Theoretical Physics, 38*, 1993–2007.

Castner, S. A., & Williams, G. V. (2007). From vice to virtue: Insights from sensitization in the nonhuman primate. *Progress in Neuro-Psychopharmacology and Biological Psychiatry, 31*, 1572–1592.

Cavanagh, P. (2011). Visual cognition. *Vision Research, 51*, 1538–1551.

Centonze, D., Siracusano, A., Calabresi, P., & Bernardi, G. (2004). The project for a scientific psychology (1895): A Freudian anticipation of LTP-memory. *Brain Research Reviews, 46*, 310–314.

Chaitin, G. J. (1982). Godel's theorem and information. *International Journal of Theoretical Physics, 22*, 941–954.

Chaplin, J., Yepez, R., & Shorvon, S. (1990). A quantitative approach to measuring the social effects of epilepsy. *Neuroepidemiology, 9*, 151–158.

Chatrian, G. E., Bergamini, L., Dondey, M., et al. (1974). A glossary of terms most commonly used by clinical electroencephalographers. *Electroencephalography and Clinical Neurophysiology, 37*, 538–548.

Chaudhury, D., Wang, L. M., & Colwell, C. S. (2005). Circadian regulation of hippocampal long-term potentiation. *Journal Biological Rhythms, 20*, 225–236.

Chiesa, A., & Serretti, A. (2010). A systematic review of neurobiological and clinical features of mindfulness meditations. *Psychological Medicine, 40*, 1239–1252.

Choi, J., Jeong, B., Rohan, M. L., Polcari, A. M., & Teicher, M. H. (2009). Preliminary evidence for white matter tract abnormalities in young adults exposed to parental verbal abuse. *Biological Psychiatry, 65*, 227–234.

Chrousos, G. P., Epstein, F., Flier, J., Reichlin, S., & Pavlou, S. (1995). The hypothalamic-pituitary-adrenal axis and immune-mediated inflammation. *New England Journal of Medicine, 332*, 1351–1362.

Chu, J., & Dill, D. (1990). Dissociative symptoms in relation to chidhood sexual and physical abuse. *American Journal of Psychiatry, 147*, 887–892.

Cojan, Y., Waber, L., Schwartz, S., Rossier, L., Forster, A., & Vuilleumier, P. (2009). The brain under self-control: Modulation of inhibitory and monitoring cortical networks during hypnotic paralysis. *Neuron, 62*, 862–875.

Cole, D. A., Nolen-Hoeksema, S., Girgus, J., & Paul, G. (2006). Stress exposure and stress generation in child and adolescent depression: A latent trait-state-error approach to longitudinal analyses. *Journal of Abnormal Psychology, 115*, 40–51.

Coleman-Mensches, K., & McGaugh, J. L. (1995). Differential involvement of the right and left amygdalae in expression of memory for aversively motivated training. *Brain Research 670*, 75–81.

Collip, D., Myin-Germeys, I., & Van Os, J. (2008). Does the concept of "sensitization" provide a plausible mechanism for the putative link between the environment and schizophrenia? *Schizophrenia Bulletin, 34*, 220–225.

Coons, P. M. (1980). Multiple personality: Diagnostic considerations. *Journal of Clinical Psychiatry, 41*, 330–336.

Coons, P. M., Milstein, V., & Marley, C. (1982). EEGstudies of two multiple personalityand a control. *Archives of General Psychiatry, 39*, 823–825.

Coons, P. M., Bowman, E. S., & Milstein, V. (1988). Multiple personality disorder. A clinical investigation of 50 cases. *Journal of Nervous and Mental Disease, 176*, 519–527.

Coons, P. M., Bowman, E. S., & Pellow, T. A. (1989). Post-traumatic aspects of the treatment of victims of sexual abuseand incest. *Psychiatric Clinics of North America, 12*, 325–327.

Cooper, J. M. (Ed.). (1997). *Plato: Complete works*. Indianapolis: Hackett. [Web source: Online Library of Liberty; Plato, The Dialogues of Plato, in 5 vols (Jowett ed.)]. http://oll.libertyfund.org/?option=com_staticxtandstaticfile=show.php%3Ftitle=166andItemid=27.

Corballis, P. M., Funnell, M. G., & Gazzaniga, M. S. (2000). An evolutionary perspective on hemispheric asymmetries. *Brain and Cognition, 43*, 112–117.

Cosmides, L., & Tooby, J. (1995). From function to structure: The role of evolutionary biology and computational theories in cognitive neuroscience. In M. Gazzaniga (Ed.), *The cognitive neurosciences* (pp. 1199–1210). Cambridge: MIT Press.

Costa, E., Davis, J. M., Dong, E., Grayson, D. R., Guidotti, A., Tremolizzo, L., et al. (2004). A GABAergic cortical deficit dominates schizophrenia pathophysiology. *Critical Reviews in Neurobiology, 16*, 1–23.

Cotterill, R. M. J. (1995). On the unity of conscious experience. *Journal of Consciousness Studies, 2*, 290–312.

Crawford, H. J. (1994). Brain dynamics and hypnosis. *International Journal of Clinical and Experimental Hypnosis, 42*, 204–232.

Crick, F. (1994). *The astonishing hypothesis. The scientific search for the soul*. London: Simon and Schuster.

Crick, F., & Clark, J. (1994). The astonishing hypothesis. An interview. *Journal of Consciousness Studies, 1*(1), 17–24.

Crick, F., & Koch, C. (1992). The problem of consciousness. *Scientific American, 267*(3), 153–159.

Crick, F., & Koch, C. (2003). A framework for consciousness. *Nature Neuroscience, 6*, 119–126.

Crick, F., & Koch, C. (1995). Are we aware of neural activity in primary visual cortex? *Nature, 375*, 121–123.

Critchley, H. D. (2002). Electrodermal responses: What happens in the brain. *Neuroscientist, 8*, 132–142.

Davidson, R. J., Kabat-Zinn, J., Schumacher, J., Rosenkranz, M., Muller, D., Santorelli, S. F., Sheridan, J. F. (2003). Alterations in brain and immune function produced by mindfulness meditation. *Psychosomatic Medicine, 65*, 564–570.

Davis, A. V., Paulsen, J. S., Heaton, R. K., & Jeste, D. V. (1995). Assessment of the semantic network in chronic schizophrenia. *Journal of the International Neuropsychological Society, 1*, 132.

De Bellis, M. D., Hall, J., Boring, A. M., Frustaci, K., & Moritz, G. (2001). A pilot longitudinal study of hippocampal volumes in pediatric maltreatment-related posttraumatic stress disorder. *Biological Psychiatry, 50*, 305–309.

De Bellis, M. D., Keshavan, M. S., Clark, D. B., Casey, B. J., Giedd, J. N., Boring, A. M., et al. (1999). Developmental traumatology part II: Brain development. *Biological Psychiatry, 45*, 1271–1284.

De Lucia, M., Bottaccio, M., Montuori, M., & Pietronero, L. (2005). Topological approach to neural complexity. *Physical Review E—Statistical, Nonlinear, and Soft Matter Physics, 71*(1), 016114.

Debiec, J., & Altemus, M. (2006). Toward a new treatment for traumatic memories. *Cerebrum*, 2–11.

Debiec, J., & LeDoux, J. E. (2006). Noradrenergic signaling in the amygdala contributes to the reconsolidation of fear memory: Treatment implications for PTSD. *Annals of the New York Academy of Sciences, 1071*, 521–524.

Debiec, J., LeDoux, J.E., & Nader, K. (2002). Cellular and systems reconsolidation in the hippocampus. *Neuron, 36*, 527–38.

Debiec, J., Doyere, V., Nader, K., & LeDoux, J. E. (2006). Directly reactivated, but not indirectly reactivated, memories undergo reconsolidation in the amygdala. *Proceedings of the National Academy of Sciences of the United States of America, 103*(9), 3428–3433.

Dehaene, S. (2009). Origins of mathematical intuitions: The case of arithmetic. *Annals of the New York Academy of Sciences, 1156*, 232–259.

Delong, R. G. (1999). Autism: New data suggest a new hypothesis. *Neurology, 52*, 911–916.

Dennett, D. (1990). Quining qualia. In W. Lycan (Ed.), *Mind and cognition* (pp. 519–548). Oxford: Blackwells.

Dennett, D. (1991). *Consciousness explained*. Boston: Little, Brown.

Derbyshire, J. (2004). *Prime obsession: Bernhard Riemann and the greatest unsolved problem in mathematics*. New York: Penguin.

Descartes, R. (1975). The passions of the soul. In E. S. Haldane & G. R. T. Ross (Eds. & Trans), *The philosophical works of Descartes* (vol. 1). London: Cambridge University Press.

Desimone, R., & Duncan, J. (1995). Neural mechanisms of selective visual attention. *Annual Review of Neuroscience, 18*, 193–222.

Desimone, R., Schein, S. J., Moran, J., & Ungerleider, L. G. (1985). Contour, color and shape analysis beyond the striate cortex. *Vision Research, 25*, 441–452.

d'Espagnat, B. (1976). *Conceptual foundations of quantum mechanics* (2nd ed.). Reading: Benjamin.

Diamond, D. M., & Rose, G. M. (1994). Stress impairs LTP and hippocampal-dependent memory. *Annals of the New York Academy of Sciences, 746*, 411–414.

Diaz, M. T., & McCarthy, G. (2007). Unconscious word processing engages a distributed network of brain regions. *Journal of Cognitive Neuroscience, 19*, 1768–1775.

Dierig, S. (1994). Extending the neuron doctrine: Carl Ludwig Schleich (1859–1922) and his reflections on neuroglia at the inception of the neural-network concept in 1894. *Trends in Neurosciences, 17*, 449–452.

Dokoumetzidis, A., Iliadin, A., & Macheras, P. (2001). Nonlinear dynamics and chaos theory: Concepts and applications relevant to pharmacodynamics. *Pharmacological Research, 18*, 415–426.

Dranovsky, A., & Hen, R. (2006). Hippocampal neurogenesis: Regulation by stress and antidepressants. *Biological Psychiatry, 59*, 1136–1143.

Duch, W. (2005). Brain-inspired conscious computing architecture. *Journal of Mind and Behavior, 26*, 1–21.

Duda, R. O., Hart, P. E., & Stork, D. G. (2000). *Pattern classification*. New York: Wiley-Interscience.

Duman, R. S., & Monteggia, L. M. (2006). A neurothrophic model for stress-related mood disorders. *Biological Psychiatry, 59*, 1116–1127.

Eack, S. M., Prasad, K. M. R., Montrose, D. M., Goradia, D. D., Dworakowski, D., Miewald, J., et al. (2008). An integrated psychobiological predictive model of emergent psychopathology among young relatives at risk for schizophrenia. *Progress in Neuro-Psychopharmacology and Biological Psychiatry, 32*, 1873–1878.

Eadie, M. J. (1998). Dreamy mental states in late 20th century neurology. *Journal of Clinical Neuroscience, 5*, 157–160.

Ebrinc, S. (2002). Hypnotizability, pain threshold and dissociative experiences. *European Neuropsychopharmacology, 12*(Suppl.), 427–428.

Eccleston, C., & Crombez, G. (1999). Pain demands attention: A cognitive-affective model of the interruptive function of pain. *Psychological Bulletin, 125*, 356–366.

Eckhorn, R., Bauer, R., Jordan, W., Brosch, M., Kruse, W., Munk, M., & Reitboeck, H. J. (1988). Coherent oscillations: A mechanism of feature linking in the visual cortex? *Biological Cybernetics, 60*, 121–130.

Edelman, G. (1989). *The remembered present*. New York: Basic Books.

Edelman, G. M. (2003). Naturalizing consciousness: A theoretical framework. *Proceedings of the National Academy of Sciences of the United States of America, 100*, 5520–5524.

Edelson, M. (1986). The convergence of psychoanalysis and neuroscience—Illusion and reality. *Contemporary Psychoanalysis, 22*, 479–519.

Einstein A. (2005). *Relativity the special and general theory*. New York: Pi Press (Originally published 1916).

Einstein, A. (1918). *Principles of research*. Berlin: Max Planck Institut.

Einstein, A., Podolsky, B., & Rosen, N. (1935). Can quantum-mechanical description of physical reality be considered complete? *Physical Review, 47*, 777–780.

Elbert, T., Ray, W. J., Kowalik, Z. J., Skinner, J. E., Graf, K. E., & Birbaumer N. (1994). Chaos and physiology: Deterministic chaos in excitable cell assemblies. *Physiological Reviews, 74*, 1–47.

Elenkov, I. J., & Chrousos, G. P. (2002). Stress hormones, proinflammatory and antiinflammatory cytokines, and autoimmunity. *Annals of the New York Academy of Sciences, 966*, 290–303.

Elenkov, I. J., Wilder, R. L., Chrousos, G. P., & Vizi, E. S. (2000). The sympathetic nerve—An integrative interface between two supersystems: The brain and the immune system. *Pharmacological Reviews, 52*, 595–638.

Ellenberger, H. F. (1970). *The discovery of the unconscious: The history and evolution of dynamic psychiatry*. New York: Basic.

El-Sherif, Y., Tesoriero, J., Hogan, M. V., & Wieraszko, A. (2003). Melatonin regulates neuronal plasticity in the hippocampus. *Journal of Neuroscience Research, 72*, 454–460.

Epstein, I. R., & Pojman, J. A. (1998). *An introduction to nonlinear chemical dynamics: Oscillations, waves, patterns, and chaos*. New York: Oxford University Press.

Erdelyi, M. H. (2004a). Comments on commentaries: Kihlstrom, Bachmann, Reingold, and Snodgrass. *Consciousness and Cognition, 13*, 430–433.

Erdelyi, M. H. (2004b). Subliminal perception and its cognates: Theory, indeterminacy, and time. *Consciousness and Cognition, 13*, 73–91.

Erdi, P. (2007). *Complexity explained*. New York: Springer.

Faber, J., & Vladyka, V. (1987). Epileptogenesis and psychosogenesis, antithesis or synthesis? *Acta Universitatis Carolinae Medica, 34*, 245–312.

Fabricius, T. (2010). The savant hypothesis: Is autism a signal-processing problem? *Medical Hypotheses, 75*, 257–265.

Faure, P., & Korn, H. (2001). Is there chaos in the brain? I. Concepts of nonlinear dynamics and methods of investigation. *Comptes rendus de l'Academie des sciences III, 324*, 773–793.

Faymonville, M. E., Boly, M., & Laureys, S. J. (2006). Functional neuroanatomy of the hypnotic state. *Journal of Physiology (Paris), 99*, 463–469.

Feigenson, L., Dehaene, S., & Spelke, E. (2004). Core systems of number. *Trends in Cognitive Sciences, 8*, 307–314.

Feinberg, I. (1978). Efference copy and corollary discharge: Implications for thinking and its disorders. *Schizophrenia Bulletin, 4*, 636–40.

Feinberg, I., & Guazzelli, M. (1999). Schizophrenia—A disorder of the corollary discharge systems that integrate the motor systems of thought with the sensory systems of consciousness. *British Journal of Psychiatry, 17*, 4196–4204.

Feldman, J. B. (2004). The neurobiology of pain, affect and hypnosis. *American Journal of Clinical Hypnosis, 46*, 187–200.

Fell, J., Fernández, G., Klaver, P., Elger, C. E., & Fries, P. (2003). Is synchronized neuronal gamma activity relevant for selective attention? *Brain Research Reviews, 42*, 265–272.

Felleman, D., & Van Essen, D. (1991). Distributed hierarchical processing in the primate visual cortex. *Cerebral Cortex, 1*, 1–47.

Ferenczi S. (1934). Gedanken über das trauma. *International Zeitschrift für Psychoanalyse, 20*, 5–12.

Ferguson, S. M., Rayport, M., Gardner, R., Kass, W., Weiner, H., & Reiser, M. F. (1969). Similarities in mental content of psychotic states, spontaneous seizures, dreams, and responses to electrical brain stimulation in patients with temporal lobe epilepsy. *Psychosomatic Medicine, 31*, 479–498.

Feynman, R. P., Leighton, R. B., & Sands, M. (2005). *The Feynman lectures on physics*. San Francisco: Pearson/Addison Wesley.

Fidelman, U. (2005). Visual search and quantum mechanics: A neuropsychological basis of Kant's creative imagination. *Journal of Mind and Behavior, 26*, 23–33.

Field, G. (1999). Astrophysics. *Reviews of Modern Physics, 71*(Suppl. 2), S33–40.

Finger, S. (1994). *Origins of neuroscience: A history of explorations into brain function*. New York: Oxford University Press.

Fischer, C. (1954). Dreams and perception: The role of preconscious and primary modes of perception and dream formation. *Journal of the American Psychoanalytic Association, 2*, 389–445.

Fishbain, D. A., Cutler, R. B., Rosomoff, H. L., & Rosomoff, R. S. (2001). Pain-determined dissociation episodes. *Pain Medicine, 2*, 216–224.

Fisher, R., van Emde, B. W., Blume, W., Elger, C., Genton, P., Lee, P., & Engel, J. (2005). Epileptic seizures and epilepsy: Definitions proposed by the International League Against Epilepsy (ILAE) and the International Bureau for Epilepsy (IBE). *Epilepsia, 46*, 470–472.

Fitzgerald, M. (2004). *Autism and creativity: Is there a link between autism in men and exceptional ability?*. East Sussex: Brunner-Routledge.

Foerstl, J. (1989). Early interest in the idiot savant. *American Journal of Psychiatry, 146*, 566.

Folina, J. (1992). *Poincaré and the philosophy of mathematics*. London: Macmillan.

Ford, J. M., & Mathalon, D. H. (2008). Neural synchrony in schizophrenia. *Schizophrenia Bulletin, 34*, 904–906.

Ford, J. M., Mathalon, D. H., Heinks, T., Kalba, S., Faustman, W. O., & Roth, W. T. (2001). Neurophysiological evidence of corollary discharge dysfunction in schizophrenia. *American Journal of Psychiatry, 158*, 2069–2071.

Ford, J. M., Gray, M., Faustman, W. O., Heinks, T. H., & Mathalon, D. H. (2005). Reduced gamma-band coherence to distorted feedback during speech when what you say is not what you hear. *International Journal of Psychophysiology, 57*, 143–150.

Ford, J. M., Gray, M., Faustman, W. O., Roach, B. J., & Mathalon, D. H. (2007). Dissecting corollary discharge dysfunction in schizophrenia. *Psychophysiology, 44*, 522–529.

Fox, R. F. (1990). Chaos, molecular fluctuations, and the correspondence limit. *Physical Review A, 41*, 2969–2976.

Frackowiak, R. S. J. (1997). *Human Brain function*. San Diego: Academic Press:

Francis, G., Grossberg, S., & Mingolla, E. (1994). Cortical dynamics of feature binding and reset: Control of visual persistence. *Vision Research, 34*, 1089–1104.

Frankel, F. H. (1996). Dissociation: The clinical realities. *American Journal of Psychiatry, 153*(Suppl.), 64–70.

Frazer, A., Brown, R., Kocsis, J., Caroff, S., Amsterdam, J., Winokur, A., Sweeney, J., & Stokes, P. (1986). Patterns of melatonin rhythms in depression. *Journal of Neural Transmission, 21*, 269–290.

Freeman, W. J. (1983). The physiological basis of mental images. *Biological Psychiatry, 18*, 1007–1025.

Freeman, W. J. (1991). The physiology of perception. *Scientific American, 264*, 78–85.

Freeman, W. J. (1999). Consciousness, intentionality, and causality. *Journal of Consciousness Studies, 6*, 143–172.

Freeman, W. J. (2000). Mesoscopics neurodynamics: From neuron to brain. *Journal of Physiology (Paris), 94*, 303–322.

Freeman, W. J. (2001). Biocomplexity: Adaptive behavior in complex stochastic dynamical systems. *BioSystems, 59*, 109–123.

Freud, S. (1895). Civilisation and its discontents. In J. Strachey (Eds., & Trans.), *The standard edition of the complete psychological works of Sigmund Freud* (vol. 21). London, Hogarth Press, 1966.

Freud, S. (1895). Project for a scientific psychology (1895). In J. Strachey (Eds., & Trans.), *The standard edition of the complete psychological works of Sigmund Freud* (vol. 1). London, Hogarth Press, 1966.

Freud, S. (1923). *The ego and the id*. New York: W.W. Norton and Company.

Freud, S. (1932/1933). New Introductory Lectures on Psycho –Analysis. In In J. Strachey (Eds., & Trans.), *The standard edition of the complete psychological works of Sigmund Freud* (vol. 22). London: Hogarth Press, 1966.

Freud, S. (1991). *The interpretation of dreams* (translated by James Strachey). London: Penguin Books.

Frieden, B. R. (2004). *Science from fisher information*. Cambridge: Cambridge University. Press.

Fries, P. (2005). A mechanism for cognitive dynamics: Neuronal communication through neuronal coherence. *Trends in Cognitive Sciences, 9*, 474–480.

Fries, P. (2009). Neuronal gamma-band synchronization as a fundamental process in cortical computation. *Annual Review of Neuroscience, 32*, 209–224.

Friston, K. J. (1997). Transients, metastability, and neuronal dynamics. *NeuroImage, 5*(2), 164–171.

Friston, K. J., Tononi, G., Sporns, O., & Edelman, G. M. (1995). Characterising the complexity of neuronal interactions. *Human Brain Mapping, 3*, 302–314.

Frith, U. (1989). *Autism*. Cambridge: Basil Blackwell.

Fröhlich, H. (1968). Long range coherence and energy storage in biological systems.' *International Journal of Quantum Chemistry, 2*, 641–664.

Fröhlich, H. (1970). Long range coherence and the actions of enzymes. *Nature, 228*, 1093.

Fröhlich, H. (1975). The extraordinary dielectric properties of biological materials and the actions of enzymes. *Proceedings of the National Academy of Sciences of the United States of America, 72*, 4211–4215.

Fuchs, E., & Schumacher, M. (1990). Psychosocial stress affects pineal function in the tree shrew (Tupaia belangeri). *Physiology and Behavior, 47*, 713–717.

Fujiwara, H., Namiki, C., Hirao, K., Miyata, J., Shimizu, M., Fukuyama, H., Sawamoto, N., Hayashi, T., & Murai, T. (2007). Anterior and posterior cingulum abnormalities and their association with psychopathology in schizophrenia: A diffusion tensor imaging study. *Schizophrenia Research, 28*, 215–222.

Gabbard, G. O. (2000). A neurobiologically informed perspective on psychotherapy. *British Journal of Psychiatry, 177*, 117–122.

Gabbard, G. O. (2007). Psychotherapy in psychiatry. *International Review of Psychiatry, 19*, 5–12.

Gabel, S. (1989). Dreams as a possible reflection of dissociated self-monitoring system. *Journal of Nervous and Mental Disease, 177*, 560–568.

Gabriel, A. (2006). Lamotrigine adjunctive treatment in resistant unipolar depression: An open, descriptive study. *Depression and Anxiety, 23*, 485–488.

Galin, D. (1974). Implications for psychiatry of left and right cerebral specialization. *Archives of General Psychiatry, 31*, 572–583.

Gallistel, C. R., & Gelman, R. (1992). Preverbal and verbal counting and computation. *Cognition, 44*, 43–74.

Gawronski, B., Hofmann, W., & Wilbur, C. J. (2006). Are "implicit" attitudes unconscious? *Consciousness and Cognition, 15*, 485–499.

Gazzaniga, M. S. (2000). Cerebral specialization and interhemispheric communication: Does the corpus callosum enable the human condition? *Brain, 123*, 1293–1326.

Gazzaniga, M. S., & Sperry, R. W. (1967). Language after section of the cerebral commissures. *Brain, 90*, 131–148.

Gelman, R., & Butterworth, B. (2005). Number and language: How are they related? *Trends in Cognitive Sciences, 9*, 6–10.

Gerdin, M. J., Masana, M. I., Rivera-Bermudez, M. A., Hudson, R. L., Earnest, D. J., Gillette. M. U., & Dubocovich, M. L. (2004). Melatonin desensitizes endogenous MT2 melatonin receptors in the rat suprachiasmatic nucleus: Relevance for defining the periods of sensitivity of the mammalian circadian clock to melatonin. *FASEB Journal, 18*, 1646–1656.

Germine, M. (1993). Information and psychopathology. *Journal of Nervous and Mental Disease, 181*, 382–387.

Ghirardi, G. C., Grassi, R., & Pearle, P. (1990). Relativistic dynamical reduction models: General framework and examples. *Foundations of Physics, 20*, 271–316.

Ghose, G. M., & Tso. (1997), p. 7.

Gilbert, C. D., & Wiesel, T. N. (1990). The influence of contextual stimuli on the orientation selectivity of cells in primary visual cortex of the cat. *Vision Research, 30*, 1689–1701.

Giles, R. (1964). *Mathematical foundations of thermodynamics*. Oxford: Pergamon.

Glansdorff, P., & Prigogine, I. (1971). *Thermodynamic theory of structure, stability and fluctuations*. London: Wiley-Interscience.

Glassman, R. B., & Malamut, B. L. (1977). Does the brain actively maintain itself? *Biosystems, 9*, 257–268.

Glenthoj, B. Y. (1995). The brain dopaminergic system. pharmacological, behavioural and electrophysiological studies. *Danish Medical Bulletin, 42*, 1–21.

Glenthoj, B. Y., & Hemmingsen, R. (1997). Dopaminergic sensitization: Implications for the pathogenesis of schizophrenia. *Progress in Neuro-Psychopharmacology and Biological Psychiatry, 21*, 23–46.

Globus, G. C., & Arpaia, J. P. (1994). Psychiatry and the new dynamics. *Biological Psychiatry, 32*, 352–364.

Gloor, P. (1990). Experiential phenomena of temporal lobe epilepsy. facts and hypotheses. *Brain, 113*, 1673–1694.

Gloor, P., Olivier, A., & Quesney, L. F. (1982). The role of the limbic system in experimental phenomena of temporal lobe epilepsy. *Annals of Neurology, 12*, 129–144.

Goddard, G. V. (1967). Development of epileptic seizures through brain stimulation at low intensity. *Nature, 214*, 1020–1021.

Goddard, G. V., McIntyre, G. C., & Leech, C. K. (1969). A permanent change in brain function resulting from daily electrical stimulation. *Experimental Neurology, 25*, 295–330.

Goebel, R., Khorram-Sefat, D., Muckli, L., Hacker, H., & Singer, W. (1998). The constructive nature of vision: Direct evidence from functional magnetic resonance imaging studies of apparent motion and motion imagery. *European Journal of Neuroscience, 10*, 1563–1573.

Goldberg, T. E., & Weinberger, D. R. (2000). Thought disorder in schizophrenia: A reappraisal of older formulations and an overview of some recent studies. *Cognitive Neuropsychiatry, 5*, 1–19.

Goldsmith, T. H. (1991). *The biological roots of human nature: Forging links between evolution and behavior*. New York: Oxford University Press.

Gonzalez-Burgos, G., & Lewis, D. A. (2008). GABA neurons and the mechanisms of network oscillations: Implications for understanding cortical dysfunction in schizophrenia. *Schizophrenia Bulletin, 34*, 944–961.

Gonzalez-Garrido, A. A., Ruiz-Sandoval, J. L., Gomez-Velazquez, F. R., de Alba, J. L., & Villasenor-Cabrera, T. (2002). Hypercalculia in savant syndrome: Central executive failure? *Archives of Medical Research, 33*, 586–589.

Good, M. I. (1993). The concept of an organic dissociative syndrome: What is the evidence? *Harvard Review of Psychiatry, 1*, 145–57.

Gorfine, T., & Zisapel, N. (2007). Melatonin and the human hippocampus, a time dependent interplay. *Journal of Pineal Research, 43*, 80–86.

Gottesman, D., & Chuang, I. (1999). Demonstrating the viability of universal quantum computation using teleportation and single-qubit operations. *Nature, 402*, 390–393.

Gottschalk, A. M., Bauer, M. S., & Whybrow, P. C. (1995). Evidence of chaotic mood variation in bipolar disorder. *Archives of General Psychiatry, 52*, 947–959.

Gove, A., Grossberg, S., & Mingolla, E. (1995). Brightness perception, illusory contours, and corticogeniculate feedback. *Visual Neuroscience, 12*, 1027–1052.

Grant, J. A., Duerden, E. G., Courtemanche, J., Cherkasova, M., Duncan, G. H., & Rainville, P. (2013). Cortical thickness, mental absorption and meditative practice: Possible implications for disorders of attention. *Biological Psychology, 92*, 275–281.

Gray, C. M. (1999). The temporal correlation hypothesis of visual feature integration: Still alive and well. *Neuron, 24*, 31–47.

Gray, J. A. (1995). The contents of consciousness: A neuropsychological conjecture. *Behavioral and Brain Sciences, 18*, 659–722.

Greaves, G. B. (1980). Multiple personality. 165 years after Mary Reynolds. *Journal of Nervous and Mental Disease, 168*, 577–596.

Grossberg, S., & Grunewald, A. (1997). Cortical synchronization and perceptual framing. *Journal of Cognitive Neuroscience, 9*, 117–132.

Grossman, A. W., Churchill, J. D., McKinney, B. C., Kodish, I. M., Otte, S. L., & Greenough, W. T. (2003). Experience effects on brain development: Possible contributions to psychopathology. *Journal of Child Psychology and Psychiatry and Allied Disciplines, 44*, 33–63.

Groth-Marnat, G. (2003). *Handbook of psychological assessment*. Hoboken: Wiley.

Guralnik, O., Levin, R., & Schmeidler, J. (1999). Dreams of personality disordered subjects. *Journal of Nervous and Mental Disease, 187*, 40–46.

Guralnik, O., Schmeidler, J., & Simeon, D. (2000). Feeling unreal: Cognitive processes in depersonalization. *American Journal of Psychiatry. 157*, 103–109.

Hagan, S., Hameroff, S. R., & Tuszynski, J. A. (2002). Quantum computation in brain microtubules: Decoherence and biological feasibility. *Physical Review E, 65*, 061901–069101.

Halgren, E., Walter, R. D., Cherlow, D. G., & Crandall, P. H. (1978). Mental phenomena evoked by electrical stimulation of the human hippocampal formation and amygdala. *Brain, 101*, 83–117.

Hamada, T., Antle, M. C., Silver, R. (2004). Temporal and spatial expression patterns of canonical clock genes and clock-controlled genes in the suprachiasmatic nucleus. *European Journal of Neuroscience, 19*, 1741–1748.

Hameroff, S. R., & Penrose, R. (1995). Orchestrated reduction of quantum coherence in brain microtubules: A model for consciousness. *Neural Network World, 5*, 793.

Hameroff, S. R., & Penrose, R. (1996). Orchestrated reduction of quantum coherence in brain microtubules: A model of consciousness. In S. Hameroff & A. Kaszniak (Eds), *Toward a science of consciousness* (pp. 507–540). Cambridge: MIT Press.

Harkness, K. L., Bruce, A. E., & Lumley, M. N. (2006). The role of childhood abuse and neglect in the sensitization to stressful life events in adolescent depression. *Journal of Abnormal Psychology, 115*, 730–741.

Haroche, S., & Raimond, J. M. (1996). Quantum computing: Dream or nightmare? *Physics Today, 8*, 51–52.

Harrison, L. G. (1993). *Kinetic theory of living pattern*. Cambridge: Cambridge University Press.

Hartmann, E. (1998). Nightmare after traumaas paradigm for all dreams: A new approach to the nature and functions of dreaming. *Psychiatry, 61*, 223–238.

Hauser, M. D., MacNeilage, P., & Ware, M. (1996). Numerical representations in primates. *Proceedings of the National Academy of Sciences of the United States of America, 93*, 1514–1517.

Havens, L. L. (1966). Pierre Janet. *Journal of Nervous and Mental Disease, 143*, 383–398.

Haynie, D. T. (2008). *Biological thermodynamics*. New York: Cambridge University Press.

Hazel, M. (2003). Einstein and Newton showed signs of autism. New Scientist. http://www.newscientist.com/article.ns?id=dn3676.

Heath, R. G. (1961/2005). Common characteristics of epilepsy and schizophrenia: Clinical observation and depth electrode studies (1961). *Epilepsy and Behavior* (2005), *6*, 633–645.

Heath, R. G. (1976). Brain function in epilepsy: Midbrain, medullary, and cerebellar interaction with the rostral forebrain. *Journal of Neurology, Neurosurgery and Psychiatry, 39*, 1037–1051.

Heavey, L., Pring, L., & Hermelín, B. (1999). A date to remember: The nature of memory in savant calendrical calculators. *Psychological Medicine, 29*, 45–60.

Heckers, S., & Konradi, C. (2002). Hippocampal neurons in schizophrenia. *Journal of Neural Transmission, 109*, 891–905.

Heim, C., Newport, D. J., Heit, S., Graham, Y. P., Wilcox, M., Bonsall, R., Miller, A. H., & Nemeroff, C. B. (2000). Pituitary-adrenal and autonomic responses to stress in women after sexual and physical abusein childhood. *JAMA: the journal of the American Medical Association, 284*, 592–7.

Heisenberg, W. (1958). *Physics and philosophy. The revolution in modern science*. New York: Harper and Row.

Hemby, S. E., Trojanowski, J. Q., & Ginsberg, S. D. (2003). Neuron-specific age-related decreases in dopamine receptor subtype mRNAs. *Journal of Comprehensive Neurology, 456*, 176–183.

Henry, G., Weingartner, H., Murphy, L. D. (1973). Influence of affective states and psychoactive drugs on verbal learning and memory. *American Journal of Psychiatry, 130*, 66–71.

Henry, J. P. (1993). Psychological and physiological responses to stress: The right hemisphere and the hypothalamo-pituitary-adrenal axis, an inquiry into problems of human bonding. *Integrative Physiology and Behavioral Sciences, 28*, 369–387.

Henry, J.P. (1997). Psychological and physiological responses to stress: The right hemisphere and the hypothalamo-pituitary-adrenal axis, an inquiry into problems of human bonding. *Acta Physiologica Scandinavica, 161*, 10–25.

Herbert, M. R., Ziegler, D. A., Deutsch, C. K., O'Brien, L. M., Kennedy, D. N., Filipek, P. A., et al. (2005). Brain asymmetries in autism and developmental language disorder: A nested whole-brain analysis. *Brain, 128*, 213–226.

Hermelin, B., & O'Connor, N. (1983). The idiot savant: Flawed genius or clever Hans? *Psychological Medicine, 13*, 479–481.
Hermelin, B., & O'Connor, N. (1986). Idiot savant calendrical calculators: Rules and regularities. *Psychological Medicine, 16*, 885–893.
Hermelin, B., & O'Connor, N. (1990a). Art and accuracy: The drawing ability of idiot-savants. *Journal of Child Psychology and Psychiatry, 31*, 217–228.
Hermelin, B., & O'Connor, N. (1990b). Factors and primes: A specific numerical ability. *Psychological Medicine, 20*, 163–169.
Hermelin, B., O'Connor, N., & Lee, S. (1987). Musical inventiveness of five idiot savants. *Psychological Medicine, 17*, 685–694.
Hermelin, B., O'Connor, N., Lee, S., & Treffert, D. (1989). Intelligence and musical improvisation. *Psychological Medicine, 19*, 447–457.
Hersch, J., Yiu-Chung, C., & Smeltzer, D. (2002). Identity shifts in temporal lobe epilepsy. *General Hospital Psychiatry, 24*, 185–187.
Hess, B., & Mikhailov, A. (1994). Self-organization in living cells. *Science, 264*, 223–224.
Hilgard, E. R. (1986). *Divided consciousness. Multiple control in human thought and action*. New York: Wiley.
Hilgetag, C. C., Kotter, R., & Young, M. P. (1999). Inter-hemispheric competition of sub-cortical structures is a crucial mechanism in paradoxical lesion effects and spatial neglect. *Progress in Brain Research, 121*, 121–141.
Hill, A. L. (1978). Mentally retarded individuals with special skills. In N. R. Eller (Ed.), *International review of research in mental retardation* (vol. 9, pp. 277–298). NewYork: Academic Pan.
Himelhoch, S., Taylor, S. F., Goldman, R. S., & Tandon, R. (1996). Frontal lobe tasks, antipsychotic medication, and schizophrenic syndromes. *Biological Psychiatry, 39*, 227–229.
Hines, M., Swan, C., Roberts, R. J., & Varney, N. R. (1995). Characteristics and mechanisms of epilepsy spectrum disorder: An explanatory model. *Applied Neuropsychology, 2*, 1–6.
Ho, E., Tsang, A., & Ho, D. (1991). An investigation of the calendar calculation ability of a Chinese calendar savant. *Journal of Autism and Developmental Disorders, 21*, 315–327.
Hoffman, R. E., & Cavus, I. (2002). Slow transcranial magnetic stimulation, long-term depotentiation, and brain hyperexcitability disorders. *American Journal of Psychiatry, 159*, 1093–1102.
Hoffman, R. E., Kirstein, L., Stopek, S., & Cicchetti, D. V. (1982). Apprehending schizophrenic dicourse: A structural analysis of the listener's task. *Brain and Language, 15*, 207–233.
Hofstadter, D. R. (1999). *Gödel, Escher, Bach: An eternal golden braid*. New York: Basic Books.
Hogan, M. V., El-Sherif, Y., & Wieraszko, A. (2001). The modulation of neuronal activity by melatonin: In vitro studies on mouse hippocampal slices. *Journal of Pineal Research, 30*, 87–96.
Hogan, R. E., & Kaiboriboon, K. (2003). The "dreamy state": John Hughlings-Jackson's ideas of epilepsy and consciousness. *American Journal of Psychiatry, 160*, 1740–1747.
Hogan, R. E., & Kaiboriboon, K. (2004). John Hughlings-Jackson's writings on the auditory aura and localization of the auditory cortex. *Epilepsia, 45*, 834–837.
Hölzel, B. K., Ott, U., Hempel, H., Hackl, A., Wolf, K., Stark, R., & Vaitl, D. (2007). Differential engagement of anterior cingulate and adjacent medial frontal cortex in adept meditators and non-meditators. *Neuroscience Letters, 421*, 16–21.
Hölzel, B. K., Carmody, J., Vangel, M., Congleton, C., Yerramsetti, S. M., Gard, T., & Lazar, S. W. (2011). Mindfulness practice leads to increases in regional brain gray matter density. *Psychiatry Research—Neuroimaging, 191*, 36–43.
Horowitz, M., Wilner, M., & Alvarez, W. (1979). Impact of event scale: A measure of subjective stress. *Psychosomatic Medicine, 41*, 209–218.
Horowitz, M. J. (1986). *Stress response syndromes* (2nd ed.). Northvale: Jason Aronson.
Horwitz, W., Kestenbaum, C., Person, E., & Jarvich, L. (1965). Identical twin "idiots savants" calendar calculators. *American Journal of Psychiatry, 121*, 1075–1079.
Horwitz, W., Deming, W., & Winter, R. (1969). A further account of the idiots savants, experts with the calendar. *American Journal of Psychiatry, 126*, 412–415.
Hotchkiss, A. P., & Harvey, P. D. (1990). Effect of distraction on communication failures in schizophrenic patients. *Clinical Research Reports, 147*, 513–515.

Howe, M. (1989a). *Fragments of genius: The strange feats of idiot savants*. London: Routledge.

Howe, M., & Smith, J. (1988). Calendar calculating in "idiots savants:" How do they do it? *British Journal of Psychology, 79*, 371–386.

Huber, M. T., Braun, H. A., & Krieg, J. C. (1999). Consequences of deterministic and random dynamics for the course of affective disorders. *Biological Psychiatry, 46*, 256–262.

Hupe, J. M., James, A. C., Payne, B. R., Lomber, S. G., Girard, P., & Bullier, J. (1998). Cortical feedback improves discrimination between figure and background by V1, V2 and V3 neurons. *Nature, 394*, 784–787.

Hurst, L., & Mulhall, D. (1988). Another calendar savant. *British Journal of Psychiatry, 152*, 274–277.

Igamberdiev, A. U. (2007). Physical limits of computation and emergence of life. *Biosystems, 90*, 340–349.

Indic, P., Schwartz, W. J., Herzog, E. D., Foley, N. C., & Antle, M. C. (2007). Modeling the behavior of coupled cellular circadian oscillators in the suprachiasmatic nucleus. *Journal of Biological Rhythms, 22*, 211–219.

Iovane, G. (2008). The set of prime numbers: Symmetries and supersymmetries of selection rules and asymptotic behaviours. *Chaos, Solitons and Fractals, 37*, 950–961.

Iovane, G., Laserra, E., & Tortoriello, F. S. (2004). Stochastic self-similar and fractal universe. *Chaos, Solitons and Fractals, 20*, 415–426.

Isenberg, N., Silbersweig, D., Engelien, A., Emmerich, S., Malavade, K., Beattie, B., & Leon, A. C. (1999). Linguistic threat activates the human amygdala. *Proceedings of the National Academy of Sciences of the United States of America, 96*, 10456–10459.

Ito, M., & Gilbert, C. D. (1999). Attention modulates contextual influences in the primary visual cortex of alert monkeys. *Neuron, 22*, 593–604.

Ito, Y., Teicher, M., Gold, C., Harper, D., Magnus, E., & Gelbard, H. (1993). Increased prevalence of electrophysiological abnormalities in children with psychological, physical and sexual abuse. *Journal of Neuropsychiatry and Clinical Neuroscience, 5*, 401–408.

Ito, Y., Teicher, M. H., Glod, C. A., & Ackerman, E. (1998). Preliminary evidence for aberrant cortical development in abused children: A quantitative EEGstudy. *Journal of Neuropsychiatry and Clinical Neuroscience, 10*, 298–307.

Ives-Deliperi, V. L., Solms, M., & Meintjes, E. M. (2011). The neural substrates of mindfulness: An fMRI investigation. *Social Neuroscience, 6*, 231–242.

Jackson, J. H. (1931). Selected writings of John Hughlins Jackson. In J. Taylor (Ed.), *On epilepsy and epileptiform convulsions* (vol. 1). London: Hodder and Stoughton.

Jacob, T. C., Moss, S. J., & Jurd, R. (2008). GABAA receptor trafficking and its role in the dynamic modulation of neuronal inhibition. *Nature Reviews Neuroscience, 9*, 331–343.

Jain, A. K., Duin, R. P. W., & Mao, J. (2000). Statistical pattern recognition: A review, *IEEE Transactions on Pattern Analysis and Machine Intelligence, 22*, 4–37.

James, I. (2003). Singular scientists. *Journal of the Royal. Society of Medicine, 96*, 36–39.

James, W. (1890). *The principles of psychology*. New York: Holt.

James, W. (1983). *The principles of psychology*. Cambridge: Harvard University Press (originally published 1890).

Jammer, M. (1989). *The conceptual development of quantum mechanics*. Los Angeles: Tomash Publishers, American Institute of Physics.

Janet, P. (1890). *L'Automatisme Psychologique*. Paris: Felix Alcan.

Jaynes, E. T. (1957). Information theory and statistical mechanics. *Physical Review, 106*(4), 620–630.

Jaynes, E. T. (1957). Information theory and statistical mechanics II. *Physical Review, 108*(2), 171–190.

Jeans, R. F. (1976). The three faces of Evelyn: A case report. I. An independently validated case of multiple personalities. *Journal of Abnormal Psychology, 85*, 249–255.

Jelicic, M., & Merckecbach, H. (2004). Traumatic stress, brain changes, and memory deficits: A critical note. *Journal of Nervous and Mental Disease, 192*, 548–553.

Jensen, F. E., & Baram, T. Z. (2000). Developmental seizures induced by common early-life insults: Short- and long-term effects on seizuresusceptibility. *Mental retardation and developmental disabilities research reviews, 6*, 253–257.
Jensen, O., Kaiser, J., & Lachaux, J. P. (2007). Human gamma-frequency oscillations associated with attention and memory. *Trends in Neurosciences, 30*, 317–324.
Jiang, Z. G., Nelson, C. S., & Allen, C. N. (1995). Melatonin activates an outward current and inhibits Ih in rat suprachiasmatic nucleus neurons. *Brain Research, 687*, 125–132.
Jindal, R. D., & Thase, M. E. (2004). Treatment of insomnia associated with clinical depression. *Sleep Medicine Reviews, 8*, 19–30.
Jobe, P. C., Dailey, J. W., & Wernicke, J. F. (1999). A noradrenergic and serotoninergic hypothesis of the linkage between epilepsy and affective disorders. *Critical Reviews in Neurobiology, 13*, 317–356.
Johannessen, L. C. (2008). Antiepileptic drugs in non-epilepsy disorders: Relations between mechanisms of action and clinical efficacy. *CNS Drugs, 22*, 27–47.
John, E. R. (2002). The neurophysics of consciousness. *Brain Research Reviews, 39*, 1–28.
Johnson, N. F. (2007). *Two's company, three is complexity: A simple guide to the science of all sciences*. Oxford: Oneworld.
Jones, E. (1953). *The life and work of Sigmund Freud*. London: Hogarth Press.
Jung, C. G. (1907). On psychophysical relations of the associative experiment. *Journal of Abnormal Psychology, 1*, 247–255.
Jung, C. G. (1909). *The psychology of dementia praecox*. New York: Journal of Nervous and Mental Disease Publishing Company (also in Collected Works of C. G. Jung 3).
Jung, C. G. (1972). *On the nature of the psyche. The structureand dynamics of the psyche* (Collected Works of C.G. Jung 8). Princeton: Princeton University Press.
Kaiser, J., & Lutzenberger, W. (2003). Induced gamma-band activity and human brain function. *Neuroscientist, 9*, 475–484.
Kalsbeek, A., Palm, I. F., La Fleur, S. E., Scheer, F. A., Perreau-Lenz, S., Ruiter, M., Kreier, F., Cailotto, C., & Buijs, R. M. (2006). SCN outputs and the hypothalamic balance of life. *Journal of Biological Rhythms, 21*, 458–469.
Kandel, E. R. (1998). A new intellectual framework for psychiatry. *American Journal of Psychiatry, 155*, 457–469.
Kandel, E. R. (1999). Biology and the future of psychoanalysis: A new intellectual framework for psychiatry revisited. *American Journal of Psychiatry, 156*(4), 505–524.
Kandel, E., Schwarz, J., & Jessel, T. (2000). *Principles of neural science*. New York: McGraw–Hill.
Kang, K., Williams, L. M., Hermens, D., & Gordon, E. (2005). Neurophysiological markers of contextual processing: The relationship between P3b and gamma synchrony and their modulation by arousal, performance and individual differences. *Cognitive Brain Research, 25*, 472–483.
Kanner, A. M. (2000). Psychosis of epilepsy: A neurologist's perspective. *Epilepsy and Behaviour, 1*, 219–227.
Kanner, A. M. (2001). The behavioural aspects of epilepsy: An overview of controversial issues. *Epilepsy and Behaviour, 2*, 8–12.
Kanner, A. M., & Balabanov, A. (2002). Depression and Epilepsy. How closely related are they? *Neurology, 58*, 827–839.
Kanner, L. (1943). Autistic disturbances of affective contact. *Nervous Child, 2*, 217–250.
Kantz, H., & Schreiber, T. (1997). *Nonlinear time series analysis*. Cambridge: Cambridge University Press.
Kanwisher, N. (2001). Neural events and perceptual awareness. *Cognition, 79*, 89–113.
Kaplan, H. I., & Sadock, B.J. (1991). *Comprehensive glossary of psychiatry and psychology*. Baltimore: Williams and Wilkins.
Karl, A., Schaefer, M., Malta, L. S., Dörfel, D., Rohleder, N., & Werner, A. (2006). A meta-analysis of structural brain abnormalities in PTSD. *Neuroscience and Biobehavioral Reviews, 30*, 1004–1031.

Karolyhazy, F. (1966). Gravitation and quantum mechanics of macroscopic objects. *IL Nuovo Cimento A, 42*, 309–342.

Karremans, J., Stroebe, W., & Claus, J. (2006). Beyond Vicary's fantasies: The impact of subliminal priming and brand choice. *Journal of Experimental Social Psychology, 42*, 792–798.

Keller, M. B. (2003). Past, present, and future directions for defining optimal treatment outcome in depression: Remission and beyond. *JAMA, 289*, 3152–3160.

Kenardy, J., Smith, A., Spence, S. H., Lilley, P. R., Newcombe, P., Dob, R., & Robinson S. (2007). Dissociation in children's trauma narratives: An exploratory investigation. *Journal of Anxiety Disorders, 21*, 456–466.

Kent, G. H., & Rosonoff, A. J. (1910). A study of associations in insanity. *American Journal of Insanity, 66/67*, 37–34; 317–390.

Kihlstrom, J. F. (1987). The cognitive unconscious. *Science, 237*, 1445–1452.

Kihlstrom, J. F. (2004). Availability, accessibility, and subliminal perception. *Consciousness and Cognition, 13*, 92–100.

Kihlstrom, J. F. (2005). Dissociative disorders. *Annual Review Clinical Psychology, 1*, 227–253.

Kilpatrick, L. A., Suyenobu, B. Y., Smith, S. R., Bueller, J. A., Goodman, T., Creswell, J. D., Luders, E., Phillips, O. R., Clark, K., Kurth, F., Toga, A. W., & Narr, K. L. (2012). Bridging the hemispheres in meditation: Thicker callosal regions and enhanced fractional anisotropy (FA) in long-term practitioners. *NeuroImage, 61*, 181–187.

Kitayama, N., Vaccarino, V., Kutner, M., Weiss, P., & Bremner, J. D. (2005). Magnetic resonance imaging (MRI) measurement of hippocampal volume in posttraumatic stress disorder: A meta-analysis. *Journal of Affective Disorders, 88*, 79–86.

Kline, P. (2000). *The handbook of psychological testing*. New York: Routledge.

Klonowski, W., Jernajczyk, W., Niedzielska, K., Rydz, A., & Stepien, R. (1999). Quantitative measure of complexity of EEG signal dynamics. *Acta Neurobiologiae Experimentalis, 59*, 315–321.

Kohut, H. (1971). *The analysis of the self: A systematic approach to the psychoanalytic treatment of narcissistic personality disorders*. New York: International Universities Press.

Korn, H., & Faure, P. (2003). Is there chaos in the brain? II. experimental evidence and related models. *Comptes Rendus Biologies, 326*, 787–840.

Koshino, H., Carpenter, P. A., Minshew, N. J., Cherkassky, V. L., Keller, T. A., & Just, M. A. (2005). Functional connectivity in an fMRI working memory task in high-functioning autism. *NeuroImage, 24*, 810–821.

Kovacs, K. J., & Elenkov, I. J. (1995). Differential dependence of ACTH secretion induced by various cytokines on the integrity of the paraventricular nucleus. *Journal of Neuroendocrinology, 7*, 15–23.

Kragh, H. (1999). *Cosmology and controversy*. Princeton: Princeton University Press.

Kraus, J. E. (2000). Sensitization phenomena in psychiatric illness: Lessons from the kindling model. *Journal of Neuropsychiatry and Clinical Neurosciences, 12*, 328–343.

Krishnamoortthy, E. S., Trimble, M. R., Sander, J. W. A. S., & Kanner, A. M. (2002). Forced normalization at the interface between epilepsyand psychiatry. *Epilepsy and Behaviour, 3*, 303–308.

Kuhn, T. (1970). *The structure of scientific revolutions*. Chicago: University of Chicago Press.

Kukleta, M., Bob, P., Brázdil, M., Roman, R., & Rektor, I. (2010). The level of frontal-temporal beta-2 band EEG synchronization distinguishes anterior cingulate cortex from other frontal regions. *Consciousness and Cognition, 19*, 879–886.

Kuyk, J., Spinhoven, P., Van Emde Boas, W., & Van Dyck, R. (1999). Dissociation in temporal lobe epilepsy and pseudo-epileptic seizure patients. *Journal of Nervous and Mental Disease, 187*, 713–720.

La Bar, K. S., Gatenby, J. C., Gore, J. C., Le Doux, J. E., & Phelps, E. A. (1998). Human amygdala activation during conditioned fear acquisition and extinction: A mixed-trial fMRI study. *Neuron, 20*, 937–945.

Lachaux, J. P., Rodriguez, E., Martinerie, J., & Varela, F. J. (1999). Measuring phase synchrony in brain signals. *Human Brain Mapping, 8*, 194–208.

Landauer, R. (1961). Irreversibility and heat generation in the computing process. *IBM Journal of Research and Development, 3*, 183–191.
Landolt, H. (1953). Some clinical electroencephalographical correlations in epileptic psychosis (twilight states). *Electroencephalography and Clinical Neurophysiology, 5*, 121.
Lane, R. D., Reiman, E. M., Axelrod, B., Yun, L. S., Holmes, A., & Schwarz, G. E. (1998). Neural correlates of levels of emotional awareness. Evidence of an interaction between emotion and attention in the anterior cingulate cortex. *Journal of Cognitive Neuroscience, 10*, 525–535.
Larmore, K., Ludwig, A. M., & Cain, R. L. (1977). Multiple personality: An objective case study. *British Journal of Psychiatry, 31*, 3540.
LaRock, E. (2006). Why neural synchrony fails to explain the unity of visual consciousness. *Behavior and Philosophy, 34*, 39–58.
Larson, J., Jessen, R. E., Uz, T., Arslan, A. D., Kurtuncu, M., Imbesi, M., & Manev, H. (2006). Impaired hippocampal long-term potentiation in melatonin MT2 receptor-deficient mice. *Neuroscience Letters, 393*, 23–26.
Laudon, M., Hyde, J. F., & Ben-Jonathan, N. (1989). Ontogeny of prolactin releasing and inhibiting activities in the posterior pituitary of male rats. *Neuroendocrinology, 50*, 644–649.
Laurikainen, K. V. (1988). *Beyond the atom: The philosophical thought of Wolfgang Pauli*. Berlin: Springer.
Lavenex, P., & Amaral, D. G. (2000). Hippocampal-neocortical interaction: A hierarchy of associativity. *Hippocampus, 10*, 420–430.
Lazar, S. W., Kerr, C. E., Wasserman, R. H., Gray, J. R., Greve, D. N., Treadway, M. T., McGarvey, M., Quinn, B.T., Dusek, J.A., Benson, H., Rauch, S.L., Moore, C.I., & Fischl, B. (2005). Meditation experience is associated with increased cortical thickness. *Neuroreport, 16*, 1893–1897.
LeDoux, J. E. (1992). Brain mechanisms of emotion and emotional learning. *Current Opinion in Neurobiology, 2*, 191–198.
LeDoux, J. E. (1993). Emotional memory systems in the brain. *Behavioural Brain Research, 58*, 69–79.
LeDoux, J. E. (1994). Emotion, memory and the brain. *Scientific American, 270*, 50–57.
Lee, J. L., Everitt, B. J., & Thomas, K. L. (2004). Independent cellular processes for hippocampal memory consolidation and reconsolidation. *Science, 304*, 839–843.
Lee, K. H., Williams, L. M., Breakspear, M., & Gordon, E. (2003). Synchronous gamma activity: A review and contribution to an integrative neuroscience model of schizophrenia. *Brain Research Reviews, 41*, 57–78.
Lee, U., Kim, S., Noh, G. J., Choi, B. M., Hwang, E., & Mashour, G. A. (2009a). The directionality and functional organization of frontoparietal connectivity during consciousness and anesthesia in humans. *Consciousness and Cognition, 18*, 1069–1078.
Lee, U., Mashour, G.A., Kim, S., Noh, G.J., & Choi, B.M. (2009b). Propofol induction reduces the capacity for neural information integration: Implications for the mechanism of consciousness and general anesthesia. *Consciousness and Cognition, 18*, 56–64.
Leroy, F., Pezard, L., Nandrino, J.L., & Beaune D. (2005). Dynamical quantification of schizophrenic speech. *Psychiatry Research, 133*, 159–171.
Leu, S. C., & Singh, V. K. (1992). Stimulation of interleukin-6 production by corticotropin-releasing factor. *Cellular Immunology, 143*, 220–227.
Levitan, H. (1980). The dreamin traumatic states. In J. M. Natterson (Ed.), *The dream in clinical practice*. New York: Jason Aronson.
Lewicki, P., Hill, T., & Czyzewski, M. (1992). Nonconscious acquisition of information. *American Psychologist, 47*, 796–802.
Li, D., & Spiegel, D. (1992). A Neural Network Model of Dissociative Disorders. *Psychiatric Annals, 22*, 144–147.
Libet, B. (1998). Do the models offer testable proposals of brain functions for conscious experience. In H. H. Jasper, L. Descarries, V. C. Costelucci, & S. Rossignol (Eds.), *Advances in neurology: Consciousness at the frontiers of neuroscience* (pp. 213–217). Philadelphia: Lippincott-Raven.

Libet, B., Alberts, W. W., Wright, E. W. Jr., Delattre, L. D., Levin, G., & Feinstein, B. (1964). Production of threshold levels of conscious sensation by electrical stimulation of human somatosensory cortex. *Journal of Neurophysiology, 27*, 546–578.

Libet, B., Alberts, W. W., Wright, E. W. Jr., & Feinstein, B. (1967). Responses of human somatosensory cortex to stimuli below threshold for conscious sensation. *Science, 158*, 1597–1600.

Libet, B., Wright, E. W. Jr, Einstein, B., & Pearl D. K. (1979). Subjective refferal of the timing for a conscious sensory experience. *Brain, 102*, 193–224.

Libet, B., Pearl, D. K., Morledge, D. A., Gleason, C. A., Hosobuchi, Y., & Barbaro, N. M. (1991). Control of the transition from sensory detection to sensory awareness in man by the duration of a thalamic stimulus: The cerebral 'time-on' factor. *Brain, 114*, 1731–1757.

Liddell, B. J., Williams, L. M., Rathjen, J., Shevrin, H., & Gordon, E. (2004). A temporal dissociation of subliminal versus supraliminal fear perception: An event-related potential study. *Journal of Cognitive Neuroscience, 16*, 479–486.

Lieb, E. H., & Yngvason, J. (1999). The physics and mathematics of the second law of thermodynamics, *Physics Reports, 314*, 1–96.

Llinas, R. (2002). *I of the Vortex*. Cambridge: Bradford Book, MIT Press.

Livengood, J. M., Zouny, C. M., & Parris, C. V. (1994). Implications of multiple personality disorder for treatment of chronic pain. *Pain Digest, 4*, 191–194.

Lorenz, K. (1935/1970). Companions as factors in the bird's environment. In R. D. Martin (Ed.), *Studies in animal and human behavior* (vol. 1, pp. 101–258). London: Methuen, & Co.

Lorenz, K. 1950. The comparative method in studying innate behaviour patterns. *Symposium of the Society for Experimental Biology, 4*, 221–268.

Lorenz, K. (1965). *Evolution and modification of behavior*. Chicago: University of Chicago Press.

Loscher, W., Jackel, R., & Czucvar, J. (1986). Is amygdala kindling in rats a model for drug-resistant partial epilepsy? *Experimental Neurology, 93*, 211–226.

Lotz, M., Vaughan, J. H., & Carson, D. A. (1988). Effect of neuropeptides on production of inflammatory cytokines by human monocytes. *Science, 241*, 1218–1221.

Luck, S. J., & Girelli, M. (1998). Electrophysiological approaches to the study of selective attention in the human brain. In R. Parasuraman (Ed.), *The attentive brain* (pp. 71–94). Cambridge: MIT Press.

Lumer, E. D., Edelman, G. M., & Tononi, G. (1997). Neural dynamics in a model of the thalamocortical system. I. layers, loops and the emergence of fast synchronous rhythms. *Cerebral Cortex, 7*, 207–227.

Lutzenberger, W., Birbaumer, N., Flor, H., Rockstroh, B., & Elbert, T. (1992). Dimensional analysis of the human EEGand inteligence. *Neuroscience Letters, 143*, 10–14.

Lutzenberger, W., Preissl, H., & Pulvermuller, F. (1995). Fractal dimension of electroencephalographic time series and underlying brain processes. *Biological Cybernetics, 73*, 477–482.

Lynch, M. A. (2004). Long-term potentiation and memory. *Physiological Reviews, 84*, 87–136.

Lynn, S. J., Maré, C., Kvaal, S., Segal, D., & Sivec, H. (1994). The hidden observer, hypnotic dreams, and age regression: Clinical implications. *American Journal of Clinical Hypnosis, 37*, 130–142.

Maaranen, P., Tanskanen, A., Honkalampi, K., Haatainen, K., Hintikka, J., & Viinamaki, H. (2005). Factors associated with pathological dissociation in the general population. *Australian and New Zealand Journal of Psychiatry, 39*, 387–394.

Mace, C. J. (1992). Hysterical conversion I: A history. *British Journal of Psychiatry, 161*, 369–377.

Mace, C. J., & Trimble, M. R. (1991). Psychosis following temporal lobe epilepsy: A report of six cases. *Journal of Neurology, Neurosurgery and Psychiatry, 54*, 639–644.

Maes, M., Yirmyia, R., Noraberg, J., Brene, S., Hibbeln, J., Perini, G, et al. (2009). The inflammatory and neurodegenerative (IandND) hypothesis of depression: Leads for future research and new drug developments in depression. *Metabolic Brain Disease, 24*, 27–53.

Maniscalco, S., Francica, F., Zaffino, R. L., Lo Gullo, N., & Plastina, F. (2008). Protecting entanglement via the quantum Zeno effect. *Physical Review Letters, 100*, 090503.

Mannheim, G., Krasnewich, D., Rumsey, J., & Hibbs, T. (1992). Calendar calculating savants (CCS): A distinct phenotype/phenomenology from autism. *Neurology, 42*(Suppl. 3) 462.

Manschreck, T. C., Maher, B. A., Rucklos, M. E., & White, M. T. (1979). The predictability of thought disordered speech in schizophrenic patients. *British Journal of Psychiatry, 134*, 595–601.

Marcel, A. J. (1983). Conscious and unconscious perception: An approach to the relations between phenomenal experience and perceptual processes. *Cognitive Psychology, 15*, 238–300.

Marker, D. (2002). *Model theory: An introduction*. New York: Springer.

Marmer, S. S. (1980a). The dreamin dissociative states. In J. M. Natterson (Ed.), *The dream in clinical practice*. New York: Jason Aronson.

Marmer, S. S. (1980b). Psychoanalysis of multiple personality. *International Journal of Psycho-Analysis, 61*, 439–459.

Maroun, M., & Richter-Levin, G. (2003). Exposure to acute stress blocks the induction of long-term potentiation of the amygdala-prefrontal cortex pathway in vivo. *Journal of Neuroscience, 23*, 4406–4409.

Marsh, L., & Rao, V. (2002). Psychiatric complications in patients with epilepsy: A review. *Epilepsy Research, 49*, 11–33.

Marshall, I. N. (1989). Consciousness and Bose-Einstein condensates. *New Ideas in Psychology, 7*, 73–83.

Mashour, G. A. (2004). Consciousness unbound: Toward a paradigm of general anesthesia. *Anesthesiology, 100*, 428–433.

Mashour, G. A. (2006). Integrating the science of consciousness and anesthesia. *Anesthesia and Analgesia, 103*, 975–982.

Mashour, G. A. (2008). Toward a general theory of unconscious processes in psychoanalysis and anesthesiology. *Journal of the American Psychoanalytic Association, 56*, 203–222.

Mason, R., & Rusak, B. (1990). Neurophysiological responses to melatonin in the SCN of short-day sensitive and refractory hamsters. *Brain Research, 533*, 15–19.

Masson, J. M. (Ed.). (1985). *The complete letters of Sigmund Freud to Wilhelm Fliess* (pp. 1887–1904). Cambridge: Harvard University Press.

Matousek, M., Wackermann, J., Palus, M., Berankova, A., Albrecht, V., & Dvorak, I. (1995). Global dimensional complexity of the EEG in healthy volunteers. *Neuropsychobiology, 31*, 47–52.

Mazur, B. (2008). When is one thing equal to some other thing? In B. Gold & R. A. Simons (Eds.), *Proof and other dilemmas: Mathematics and philosophy*. Washington, DC: Mathematical Association of America.

Mc Clelland, J. L., Rumelhart, D. E., & The PDP Research Group. (1986). *Parallel distributed processing I, II*. Cambridge: MIT Press.

McCarley, R. W., Niznikiewicz, M. A., Salisbury, D. F., Nestor, P. G., O'Donnell, B. F., Hirayasu, Y., Grunze, H., Greene, R. W., & Shenton, M. E. (1999). Cognitive dysfunction in schizophrenia: Unifying basic research and clinical aspects. *European Archives of Psychiatry and Clinical Neuroscience, 249*(Suppl 4), 69–82.

McClelland, J. L. (1998). Complementary learning systems in the brain: A connectionist approach to explicit and implicit cognition and memory. *Annals of the New York Academy of Sciences, 843*, 153–169.

McFadden, I. J. (1992). Multiple personality disorder becoming manifest after persistence of a chronic pain syndrome. *Clinical Journal of Pain, 8*, 63.

McFadden, I. J., & Woitalla, V. F. (1993). Differing reports of pain perception by different personalities in a patient with chronic pain and multiple personality disorder. *Pain, 55*, 379–382.

McGaugh, J. L. (2000). Memory—A century of consolidation. *Science, 287*, 248–251.

McGrath, J. (1991). Ordering thoughts on thought disorder. *British Journal of Psychiatry, 158*, 307–316.

McNamara, J. O., Morrisett, R., & Nadler, J. V. (1992). Recent advances in understanding mechanisms of the kindling model. *Advances in Neurology, 57*, 555–560.

Meador, K. J., Ray, P. G., Echauz, J. R., Loring, D. W., & Vachtsevanos, G. J. (2005). Gamma coherence and conscious perception. *Neurology, 59*, 1847–1854.

Meares, R. (1999). The contribution of Hughlings Jackson to an understanding of dissociation. *American Journal of Psychiatry, 156*, 1850–1855.

Meduna, L. (1934). Über experimentelle Campherepilepsie. *Archiv für Psychiatrie, 102*, 333–339.
Meduna, L. (1935). Versuche über die biologische Beeinflussung des Ablaufes der Schizophrenie. Campher- und Cardiazolkräpfe. *Zeitschrift für Neurologie und Psychiatrie, 152*, 235–262.
Melancon, G., & Joanette, Y. (2000). Chaos, brain and cognition: Toward a nonlinear order? *Brain and Cognition, 42*, 33–36.
Melloni, L., Molina, C., Pena, M., Torres, D., Singer, W., & Rodriguez, E. (2007). Synchronization of neural activity across cortical areas correlates with conscious perception. *Journal of Neuroscience, 27*, 2858–2865.
Merikle, P. M., Smilek, D., & Eastwood, J. D. (2001). Perception without awareness: Perspectives from cognitive psychology. *Cognition, 79*, 115–134.
Merskey, H. (1992). The manufacture of personalities. The production of multiple personality disorders. *British Journal of Psychiatry, 160*, 327–340.
Mesulam, M. M. (1981). Dissociative states with abnormal temporal lobe EEG. *Archives of Neurology, 38*, 176–181.
Mesulam, M. M. (1998). From sensation to cognition. *Brain, 121*, 1013–1052.
Meyer, U., Kruhoffer, M., Flugge, G., & Fuchs, E. (1998). Cloning of glucocorticoid receptor and mineralocorticoid receptor cDNA and gene expression in the central nervous system of the tree shrew (Tupaia belangeri). *Molecular Brain Research, 55*, 243–253.
Meyer-Lindenberg, A. (1996). The evolution of complexity in human brain development: An EEG study. *Electroencephalography and Clinical Neurophysiology, 99*, 405–411.
Meyer-Lindenberg, A., Zeman, U., Hajak, G., Cohen, L., & Berman, K. F. (2002). Transitions between dynamical states of differing stability in the human brain. *Proceedings of the National Academy of Sciences of the United States of America, 99*, 10948–10953.
Micheloyannis, S., Flitzanis, N., Papanikolau, E., Bourkas, M., Terzakis, D., Arvanitis, S., & Stam, C. J. (1998). Usefulness of non-linear EEG analysis. *Acta Neurologica Scandinavica, 97*, 13–19.
Milin, J., Demajo, M., & Todorovic, V. (1996). Rat pinealocyte reactive response to a long-term stress inducement. *Neuroscience, 73*, 845–854.
Miller, B. L., Ponton, M., Benson, D. F., Cummings, J. L., & Mena, I. (1996). Enhanced artistic creativity with temporal lobe degeneration. *Lancet, 348*, 1744–1745.
Miller, B. L., Cummings, J., Mishkin, F., Boone, K., Prince, F., Ponton, M., et al. (1998). Emergence of artistic talent in frontotemporal dementia. *Neurology, 51*, 978–982.
Miller, G. E., Chen, E., & Zhou, E. S. (2007). If it goes up, must it come down? Chronic stress and the hypothalamic-pituitary-adrenocortical axis in humans. *Psychological Bulletin, 133*, 25–45.
Miller, G. L. (1976). Riemann's hypothesis and tests for primality. *Journal of Computer and System Sciences, 13*, 300–317.
Miller, L. K. (1987). Sensitivity to tonal structure in a developmentally disabled musical savant. *American Journal of Mental Deficiency, 91*, 467–471.
Miller, L. K. (1989). *Musical savants: Exceptional skill in the mentally retarded.* Hillsdale: Erlbaum.
Miller, L. K. (1992). Diderot reconsidered: Visual impairment and auditory compensation. *Journal of Visual Impairment and Blindness, 86*, 206–211.
Miller, L. K. (1994). The "savant syndrome:" Exceptional skill and mental retardation. In N. Colangelo, S. Assouline, & D. Ambroson (Eds.), *Talent development* (pp. 215–239). Dayton: Ohio University Press.
Miller, L. K. (1995). Sensitivity to sequential structure in musical savants. *American Journal of Mental Retardation, 99*, 391–399.
Minshew, N. J., & Hobson, J. A. (2008). Sensory sensitivities and performance on sensory perceptual tasks in high-functioning individuals with autism. *Journal of Autism and Developmental Disorders, 38*, 1485–1498.
Mizuno, K., & Giese, K. P. (2005). Hippocampus-dependent memory formation: Do memory type-specific mechanisms exist? *Journal of Pharmacological Sciences, 98*, 191–197.
Möhler, H. (2006). GABAA receptor diversity and pharmacology. *Cell and Tissue Research, 326*, 505–516.

Molle, M., Albrecht, C., & Mashall, L. (1997). Adrenocorticotropin widens the focus of attention in humans. A nonlinear electroencephalographic analysis. *Psychosomatic Medicine, 59*, 497–502.

Molle, M., Marshall, L., & Pietrowsky, R. (1996). Enhanced dynamic complexity in the human EEG during creative thinking. *Neuroscience Letters, 208*, 1–4.

Molle, M., Marshall, L., & Pietrowsky, R., (1995). Dimensional complexity of the EEG indicates a right fronto-cortical locus of attentional control. *Journal of Psychophysiology, 9*, 45–55.

Monastyrsky, M. (1999). *Riemann, topology and physics*. Boston: Birkhauser.

Monroe, R. R. (1982). Limbic ictus and atypical psychoses. *Journal of Nervous and Mental Disease, 170*, 711–716.

Monroe, S. M., & Harkness, K. L. (2005). Life stress, the "kindling" hypothesis, and the recurrence of depression: Considerations from a life stress perspective. *Psychological Review, 112*, 417–445.

Monroe, S. M., Slavich, G. M., Torres, L. D., & Gotlib, I. H. (2007). Major life events and major chronic difficulties are differentially associated with history of major depressive episodes. *Journal of Abnormal Psychology, 116*, 116–124.

Moran, L. J., Mefferd, R. B. Jr., & Kimble, J. P. Jr. (1964). Idiodynamic sets in word association. *Psychological Monographs, 78*, 1–22.

Moriarty, J., Ring, H., & Robertson, M. (1993). An idiot savant calendar calculator with Gilles de la Tourette syndrome: Implications for an understanding of savant syndrome. *Psychological Medicine, 23*, 1019–1021.

Morishima, A., & Brown, L. (1977). A case report of the artistic talent of an autistic idiot savant. *Mental Retardation, 15*, 33–37.

Morishita, S. (2009). Clonazepam as a therapeutic adjunct to improve the management of depression: A brief review. *Human Psychopharmacology, 24*, 191–198.

Morris, J. S., Ohman, A., & Dolan, R. J. (1999). A subcortical pathway to the right amygdala mediating "unseen" fear. *Proceedings of the National Academy of Sciences of the United States of America 96*, 1680–1685.

Morrison, A.P., Frame, L., & Larkin, W. (2003). Relationships between traumaand psychosis: A review and integration. *British Journal of Clinical Psychology, 42*, 331–353.

Mossner, R., Daniel, S., Albert, D., Heils, A., Okladnova, O., Schmitt, A., & Lesch, K.P. (2000). Serotonin transporter function is modulated by brain derived neurotrophic factor (BDNF) but not nerve growth factor (NGF). *Neurochemistry International, 36*, 197–202.

Mould, R. A. (1995). The inside observer in quantum mechanics. *Foundations of Physics, 25*, 1621–1629.

Mould, R. A. (1998). Consciousness and quantum mechanics. *Foundations of Physics, 28*, 1703–1718.

Mulckhuyse, M., & Theeuwes, J. (2010). Unconscious attentional orienting to exogenous cues: A review of the literature. *Acta Psychologica (Amst), 134*, 299–309.

Murray, A. L. (2010). Can the existence of highly accessible concrete representations explain savant skills? Some insights from synaesthesia. *Medical Hypotheses, 74*, 1006–1012.

Musshoff, U., Riewenherm, D., Berger, E., Fauteck, J. D., & Speckmann, E. J. (2002). Melatonin receptors in rat hippocampus: Molecular and functional investigations. *Hippocampus, 12*, 165–173.

Nadel, L. (1994). Hippocampus, space, and relations. *Behavioural and Brain Sciences, 17*, 490–491.

Nadel, L., & Jacobs, W. J. (1998). Traumatic memory is special. *Current Directions in Psychological Science, 7*, 154–157.

Nadel, L., & Moscovitch, M. (2001). The hippocampal complex and long-term memory revisited. *Trends in Cognitive Sciences, 5*, 228–230.

Nadel, L., Samsonovich, A., Ryan, L., & Moscovitch, M. (2000) Multiple trace theory of human memory: Computational, neuroimaging, and neuropsychological results. *Hippocampus 10*, 352–368.

Nagel, E., & Newman, J. (2002). *Gödel's proof.* New York: New York University Press.

Narita, M., Aoki, K., Takagi, M., Yajima, Y., & Suzuki, T. (2003). Implication of brain derived neurotrophic factor in the release of dopamine and dopamine-related behaviors induced by amphetamine. *Neuroscience, 119*, 767–775.

Nasrallah, H. A. (1985). The unintegrated right cerebral hemispheric consciousness as alien intruder: A possible mechanism for schneiderian delusions in schizophrenia. *Comprehensive Psychiatry, 26*, 273–282.

Neligan, A, & Shorvon, S. D. (2010). Frequency and prognosis of convulsive status epilepticus of different causes: A systematic review. *Archives of Neurology, 67*, 931–940.

Nelson, E., & Pribor, E. (1993). A calendar savant with autism and Tourette syndrome: Response to treatment and thoughts on the interrelationships of these conditions. *Annals of Clinical Psychiatry, 5*, 135–140.

Nemiah, J. C. (1981). Dissociative disorders. In A. M. Freeman & H. I. Kaplan (Eds.), *Comprehensive textbook of psychiatry* (3rd ed.). Baltimore: Williams and Wilkins.

Newton, I. (1686/1974). *Mathematical principles of natural philosophy*. Berkeley: University of California Press.

Nicolis, G., & Prigogine, I. (1977). *Self-organization in nonequilibrium systems*. New York: Wiley.

Nieder, A. (2005). Counting on neurons: The neurobiology of numerical competence. *Nature Reviews Neuroscience, 6*, 177–190.

Nielsen, M. A., & Chuang I. L. (2000). *Quantum computation and quantum information*. Cambridge: Cambridge University Press.

Nijenhuis, E. R. S., Spinhoven, P., Vanderlinden, J., van Dyck, R., & van der Hart, O. (1998). Somatoform dissociative symptoms as related to animal defensive reactions to predatory imminence and injury. *Journal of Abnormal Psychology, 107*, 63–73.

Nijenhuis, E. R., Spinhoven, P., van Dyck, R., van der Hart, O., & Vanderlinden, J. (1996). The development and psychometric characteristics of the Somatoform Dissociation Questionnaire (SDQ-20). *Journal of Nervous and Mental Disease, 184*, 688–694.

Nijenhuis, E. R., van der Hart, O., Kruger, K., & Steele, K. (2004). Somatoform dissociation, reported abuse and animal defence-like reactions. *Australian and New Zealand Journal of Psychiatry, 38*, 678–686.

Nijenhuis, E. R. S. (2000). Somatoform dissociation: Major symptoms of dissociative disorders. *Journal of Trauma and Dissociation, 1*, 7–32.

Nomura, Y., Nagao, Y., Kimura, K., Hachimori, K., & Segawa, M. (2010). Epilepsy in autism: A pathophysiological consideration. *Brain and Development, 32*, 799–804.

Noreika, V., Falter, C. M., Arstila, V., Wearden, J. H., & Kallio, S. (2012). Perception of short time scale intervals in a hypnotic virtuoso. *International Journal of Clinical and Experimental Hypnosis, 60*, 318–337.

O'Brien, M. (1985). The diagnosis of multiple personality syndromes: Overt, covert, and latent. *Comprehensive Therapy, 11*, 59–66.

O'Connor, N., & Hermelin, B. (1984). Idiot savant calendrical calculators: Math or memory? *Psychological Medicine, 14*, 801–806.

O'Connor, N., & Hermelin, B. (1987a). Visual and graphic abilities of the idiot savant artist. *Psychological Medicine, 17*, 79–90.

O'Connor, N., & Hermelin, B. (1987b). Visual memory and motor programmes: Their use by idiot-savant artists and controls. *British Journal of Psychology, 78*, 307–323.

O'Connor, N., & Hermelin, B. (1989). The memory structure of autistic idiot savant mnemonists. *British Journal of Psychology, 80*, 97–111.

O'Connor, N., & Hermelin, B. (1990). The recognition failure and graphic success of idiot-savant artists. *Journal of Child Psychology and Psychiatry, 31*, 203–215.

O'Connor, N., & Hermelin, B. (1991a). A specific linguistic ability. *American Journal of Mental Retardation, 95*, 673–681.

O'Connor, N., & Hermelin, B. (1991b). Talents and pre-occupations in idiots savants. *Psychological Medicine, 21*, 959–964.

O'Connor, N., & Hermelin, B. (1992). Do young calendrical calculators improve with age? *Journal of Child Psychology and Psychiatry, 33*, 907–912.

O'Keefe, L. P., & Movshon, J. A. (1998). Processing of first- and second-order motion signals by neurons in area MT of the macaque monkey. Visual *Neuroscience, 15*, 305–317.

Olypher, A. V., Klement, D., & Fenton, A. A., 2006. Cognitive disorganization in hippocampus: A physiological model of the disorganization in psychosis. *Journal of Neuroscience, 26*, 158–168.

Orbach, I., Mikulincer, M., King, R., Cohen, D., & Stein, D. (1997). Thresholds and tolerance of physical pain in suicidal and nonsuicidal adolescents. *Journal of Consulting and Clinical Psychology, 65*, 646–652.

Orr, S. P., & Roth, W. T. (2000). Psychophysiological assessment: Clinical applications for PTSD. *Journal of Affective Disorders, 61*, 225–240.

Ott, E. (2002). *Chaos in dynamical systems*. Cambridge: Cambridge University Press.

Overton, D. A. (1978). Major theories of state dependent learning. In B. T. Ho, D. V. Richards, & D. L. Chute (Eds.), *Drug discrimination and state dependent learning*. London: Academic Press.

Ozcan, M., Yilmaz, B., & Carpenter, D. O. (2006). Effects of melatonin on synaptic transmission and long-term potentiation in two areas of mouse hippocampus. *Brain Research, 1111*, 90–94.

Pacchierotti, C., Iapichino, S., Bossini, L., Pieraccini, F., & Castrogiovanni, P. (2001). Melatonin in psychiatric disorders: A review on the melatonin involvement in psychiatry. *Frontiers in Neuroendocrinology, 22*, 18–32.

Packard, R. C., & Brown, F. (1986). Multiple headaches in a case of multiple personality disorder. *Headache, 26*, 99–102.

Pakalnis, A., Drake, M. E., Kuruvilla, J., & Blake, J. K. (1988). Forced normalization. *Archives of Neurology, 45*, 139.

Panksepp, J., & Solms, M. (2012). What is neuropsychoanalysis? clinically relevant studies of the minded brain. *Trends in Cognitive Sciences, 16*(1), 6–8.

Papoulis, A. (1991). *Probability, random variables, and stochastic processes*. New York: McGraw-Hill.

Parmeggiani, A., Barcia, G., Posar, A., Raimondi, E., Santucci, M., & Scaduto, M. C. (2010). Epilepsy and EEG paroxysmal abnormalities in autism spectrum disorders. *Brain and Development, 32*, 783–789.

Pashler, H. (1998). *The psychology of attention*. Cambridge: MIT Press.

Patti, P., & Lupinetti, L. (1993). Brief report: Implications of hyperlexia in an autistic savant. *Journal of Autism and Developmental Disorders, 23*, 397–405.

Pauli, W. (1958). *Theory of relativity*. New York: Dover Publications, Inc.

Paulsen, J. S., Romero, R., Chan, A., Davis, A. V., Heaton, R. K., & Jeste, D. V. (1996). Impairment of the semantic network in schizophrenia. *Psychiatry Research, 63*, 109–121.

Paulus, M. P., & Braff, D. L. (2003). Chaos and Schizophrenia: Does the method fit the madness? *Biological Psychiatry, 53*, 3–11.

Paus, T. (2001). Primate anterior cingulate cortex: Where motor control, drive and cognition interface. *Nature Reviews Neuroscience, 2*, 417–424.

Payne, J. D., Jackson, E. D., Ryan, L., Hoscheidt, S., Jacobs, J. W., & Nadel, L. (2006). The impact of stress on neutral and emotional aspects of episodic memory. *Memory, 14*, 1–16.

Pearce, S., Isherwood, S., Hrouda, D., Richardson, P., Erskine, A., Skinner, J. (1990). Memory and pain; tests of mood congruity and state dependent learningin experimentally induced and clinical pain. *Pain, 43*, 187–193.

Pediaditakis, N. (1992). Deterministic non-linear chaosin brainfunction and borderline psychopathological phenomena. *Medical Hypotheses, 39*, 67–72.

Peled, A. (1999). Multiple constraint organization in the brain: A theory for schizophrenia. *Brain Research Bulletin, 49*, 245–250.

Pellionisz, A., & Llinas, R. (1982). Space-time representation in the brain: The cerebellum as a predictive space-time metric tensor. *Neuroscience, 7*, 2949–2970.

Penfield, W. (1967). *The excitable cortex in conscious man*. Liverpool: Liverpool University Press.

Penfield, W., & Perot, P. (1963). The brain's record of auditory and visual experience: A final summary and discussion. *Brain: A Journal of Neurology, 86*, 595–696.

Penrose, R. (1989). *The Emperor's new mind: Concerning computers, minds, and the laws of physics*. Oxford: Oxford University Press.

Penrose, R. (1994). *Shadows of the mind: An approach to the missing science of consciousness*. Oxford: Oxford University Press.

Penrose, R. (1997). *The large the small and the human mind*. Cambridge: The Press Syndicate of the University of Cambridge.

Penrose, R. (2001). Consciousness, the brain, and spacetime geometry: An addendum. *Annals of the New York Academy of Sciences, 929*, 105–110.

Penrose, R. (2002). Gravitational collapse: The role of general relativity. *General Relativity and Gravitation, 34*, 1141–1165.

Penrose, R. (2004). *The road to reality: A complete guide to the laws of the universe*. London: Jonathan Cape.

Penrose, R., & Hameroff, S. R. (1995). What "gaps"? Reply to Grush and Churchland. *Journal of Consciousness Studies, 2*, 98–111.

Pepperberg, I. M. (1994). Numerical competence in an African gray parrot (Psittacus erithacus). *Journal of Comparative Psychology, 108*, 36–44.

Perrine, K. R. (1991). Psychopathology in epilepsy. *Seminars in Neurology, 11*, 175–181.

Perrot, P. (1998). *A to Z of thermodynamics*. Oxford: Oxford University Press.

Pessoa, L. (2005). To what extent are emotional visual stimuli processed without attention and awareness? *Current Opinion in Neurobiology, 15*, 188–196.

Peterson, I. (1993). *Newton's clock: Chaos in the solar system*. New York: W.H. Freeman.

Petitot, J. (2003). The neurogeometry of pinwheels as a sub-Riemannian contact structure. *Journal of Physiology Paris, 97*, 265–309.

Pevet, P., Agez, L., Bothorel, B., Saboureau, M., Gauer, F., Laurent, V., & Masson-Pevet, M. (2006). Melatonin in the multi-oscillatory mammalian circadian world. *Chronobiology International, 23*, 39–51.

Phillips, R. G., & LeDoux, J. E. (1992). Differential contribution of amygdala and hippocampus to cued and contextual fear conditioning. *Behavioural Neuroscience, 106*, 274–285.

Phillips, T. J., Roberts, A. J., & Lessov, C. N. (1997). Behavioral sensitization to ethanol: Genetics and the effects of stress. *Pharmacology Biochemistry and Behavior, 57*, 487–493.

Picton, T. W, & Stuss. D. T. (1994). Neurobiology of conscious experience. *Current Biology, 4*, 256–265.

Plenio, M. B., & Vitelli, V. (2001). The physics of forgetting: Landauer's erasure principle and information theory. *Contemporary Physics, 42*(1), 25–60.

Plotnitsky, A. (2009). Bernhard Riemann's conceptual mathematics and the idea of space. *Configurations, 17*, 105–130.

Poetzl, O. (1960). The relationship between experimentally induced dream images and indirect vision. *Psychological Issues Monograph, 2*, 46–106.

Poincaré, H. (1998). *Science and method*. London: Thomas Nelson and Sons.

Posner, M. I., Rothbart, M. K., Sheese, B. E., & Tang, Y. (2007). The anterior cingulate gyrus and the mechanism of self-regulation. *Cognitive, Affective, and Behavioral Neuroscience, 7*, 391–395.

Post, R. M. (1992). Transduction of psychosocial stress into the neurobiology of recurrent affective disorder. *American Journal of Psychiatry, 149*, 999–1010.

Post, R. M., & Weiss, S. R. B. (1996). A speculative model of affective illness cyclicity based on patterns of drug tolerance observed in amygdala-kindled seizures. *Molecular Neurobiology, 13*, 33–60.

Post, R. M., Weiss, S. R. B., Smith, M., Li, H., & McCann, U. (1997). Kindling versus quenching. Implications for the evolution and treatment of posttraumatic stress disorder. *Annals of the New York Academy of Sciences, 821*, 285–295.

Post, R. M., Weiss, S. R., & Pert, A. (1988). Implications of behavioral sensitization and kindling for stress-induced behavioral change. *Advances in Experimental Medicine and Biology, 245*, 441–463.

Post, R. M., & Weiss, R. B. (1998). Sensitization and kindling phenomena in mood, anxiety, and obsessive-compulsive disorders: The role of serotonergic mechanisms in illness progression. *Biological Psychiatry, 44*, 193–206.

Post, R. M., Weis, S. R., & Smith, M. A. (1995). Sensitization and kindling. In M. J. Friedman, D. S. Charney, & A. Y. Deutch (Eds.), *Neurobiological and clinical consequences of stress: From normal adaptation to posttraumatic stress disorder*. Philadelphia: Lipincott-Raven.

Poulet, J. F., & Hedwig, B. (2007). New insights into corollary discharges mediated by identified neural pathways. *Trends in Neurosciences, 30*, 14–21.

Pribram, K. H., & Gill, M. M. (1976). *Freud's "project" reassessed*. London: Hutchinson.

Prigogine, I., & Stengers, I. (1984). *Order out of chaos: Man's new dialogue with nature*. New York: Bantam.

Pritchard, W. S., & Duke, D. (1995). Measuring "chaos" in the brain: A tutorial review of EEG dimension estimation. *Brain and Cognition, 27*, 353–397.

Puente, A., & Heller, S. (1987). Neuropsychological analysis of an idiot savant. *Journal of Clinical and Experimental Neuropsychology, 9*, 267–268.

Pulver, S. E. (2003). On the astonishing clinical irrelevance of neuroscience. *Journal of the American Psychoanalytic Association, 51*, 755–772.

Putnam, F. (1997). *Dissociation in children and adolescents. A developmental perspective*. London: The Guilford Press.

Putnam, F. W. (1989). *Diagnosis and treatment multiple personality disorder*. New York: The Guilford Press.

Putnam, F. W. (1992). Using hypnosisfor therapeutic abreactions. *Psychiatric Medicine, 10*, 51–65.

Putnam, F. W. (1995). Traumatic stress and pathological dissociation. *Annals of the New York Academy of Sciences, 771*, 708–715.

Putnam, F. W., Guroff, J. J., Silberman, E. K., & Post, B. L. (1986). The clinical phenomenology of multiple personality disorder. Review of 100 recent cases. *Journal of Clinical Psychiatry, 47*, 285–293.

Rainville, P., Duncan, G. H., Price, D. D., Carrier, B., & Bushnell, M. C. (1997). Pain affect encoded in human anterior cingulate but not somatosensory cortex. *Science, 277*, 968–971.

Rainville, P., Hofbauer, R. K., Bushnell, M. C., Duncan, G. H., & Price, D. D. (2002). Hypnosis modulates activity in brain structures involved in the regulation of consciousness. *Journal of Cognitive Neuroscience, 14*, 887–901.

Raizada, R. D. S., & Grossberg, S. (2001). Context-sensitive binding by the laminar circuits of V1 and V2: A unified model of perceptual grouping, attention, and orientation contrast. *Visual Cognition, 8*, 431–466.

Ramachandran, V. S., & Hirstein, W. (1997). Three laws of qualia: What neurology tells us about the biological functions of consciousness. *Journal of Consciousness Studies, 4*, 429–457.

Ramachandran, V. S., & Hubbard, E. M. (2001). Synaesthesia—A window into perception, thought and language. *Journal of Consciousness Studies, 8*, 3–34.

Rattiner, L. M., Davis, M., & Ressler, K. J. (2005). Brain-derived neurotrophic factor in amygdala-dependent learning. *Neuroscientist, 11*, 323–333.

Read, J., Perry, B. D., Moskowitz, A., & Connolly, J. (2001). The contribution of early traumatic events to schizophreniain some patients: A traumagenic neurodevelopmental model. *Psychiatry, 64*, 319–345.

Reanault, B., Signoret, J. L., Debruille, B., Breton, F., & Bolgert, F. (1989). Brain potentials reveal covert facial recognition in prosopagnosia. *Neuropsychologia, 27*, 905–912.

Rees, G., Kreiman, G., & Koch, C. (2002). Neural correlates of consciousness in humans. *Nature Reviews. Neuroscience, 3*, 261–270.

Rees, G., Russell, C., Frith, C. D., & Driver, J. (1999). Inattentional blindness versus inattentional amnesia for fixated but ignored words. *Science, 286*, 2504–2507.

Reinders, A. A. T. S., Nijenhuis, E. R. S., Paans, A. M. J., Korf, J., Willemsen, A. T. M., & den Boer, J. A. (2003). One brain, two selves. *Neuroimage, 20*, 2119–2125.

Reingold, E. M. (2004). Unconscious perception: Assumptions and interpretive difficulties. *Consciousness and Cognition, 13*, 117–122.

Rickeport, M. M. (1992). The interface between multiple personality, spirit mediumship and hypnosis. *American Journal of Clinical Hypnosis, 34*, 168–177.

Riemann, B. (1873). On the hypotheses which lie at the foundation of geometry. *Nature, 8*, 183–199.
Rife, D. C, & Snyder, L. H (1931). Studies in human inheritance VI: A genetic refutation of the principles of refutation of the principles of "behavioristic" psychology. *Human Biology, 3*, 547–559.
Riklan, M., Kabat, C., & Cooper, I. S. (1976). Psychological effects of short term cerebellar stimulation in epilepsy. *Journal of Nervous and Mental Disease, 162*, 282–290.
Riklan, M., Cullinan, T., & Cooper, I. S. (1977). Tension reduction and alerting in man following chronic cerebellar stimulation for the relief of spasticity or intractable seizures. *Journal of Nervous and Mental Disease, 164*, 176–181.
Rimland, B. (1978). Savant capabilities of autistic children and their cognitive implications. In G. Serban (Ed.), *Cognitive defects in the development of mental illness* (pp. 44–63). New York: Brunner/Mazel.
Roberts, R. J., Gorman, L. L., Lee, G. P., Hines, M. E., Richardson, E. D., Riggle, T. A., & Varney, N. R. (1992). The phenomenology of multiple partial seizurelike symptoms without stereotyped spells: An epilepsyspectrum disorder? *Epilepsy Research, 13*, 167–177.
Röder, C. H., Michal, M., Overbeck, G., van de Ven, V. G., Linden, D. E. (2007). Pain response in depersonalization: A functional imaging study using hypnosis in healthy subjects. *Psychotherapy and Psychosomatics, 76*, 115–121.
Rodriguez, R., Kallenbach, U., Singer, W., & Munk, M. H. (2004). Short- and long-term effects of cholinergic modulation on gamma oscillations and response synchronization in the visual cortex. *Journal of Neuroscience, 24*, 10369–10378.
Roediger, H. L. (1990). Implicit memory: Retention without remembering. *American Psychologist, 45*, 1043–1056.
Rofe, Y. (2008). Does repression exist? Memory, pathogenic, unconscious and clinical evidence. *Review of General Psychology, 12*, 63–85.
Roland, P. E. (1993). *Brain activation.* New York: Wiley-Liss.
Roth, T. L., & Sweatt, J. D. (2011). Epigenetic marking of the BDNF gene by early-life adverse experiences. *Hormones and Behavior, 59*, 315–320.
Rothstein, J. (1951). Information, measurement, and quantum mechanics. *Science, 114*, 171–175.
Rubin, N. (2003). Binocular rivalry and perceptual multi-stability. *Trends in Neurosciences, 26*, 289–291.
Rucker, R. (1982). *Infinity and the mind: The science and philosophy of the infinite*. Princeton: Princeton University Press.
Rueda, M. R., Posner, M. I., & Rothbart, M. K. (2004). Attentional control and self-regulation. In R. F. Baumeister & K. D. Vohs (Eds.), *Handbook of self-regulation: Research, theory, and applications* (pp. 283–300). New York: Guilford.
Ruel, J. M. H. M., & de Kloet, E. R. (1985). Two receptor systems for corticosterone in rat brain: Microdistribution and differential occupation. *Endocrinology, 117*, 2505–2512.
Rumsey, J., Mannheim, G., Aquino, T., Gordon, C., & Hibbs, E. (1992). Neuropsychological characteristics of calendar calculating savants with pervasive developmental disorders. *Clinical Neuropsychologist, 6*, 320.
Russr, M. J., Shearin, E. N., Clarkin, J. E., Harrison, K., & Hull, J. W. (1993). Subtypes of self-injurious patients with borderline personality disorder. *American Journal of Psychiatry, 150*, 1869–1877.
Rutter, J., Reick, M., & McKnight, S. L. (2002). Metabolism and the control of circadian rhythms. *Annual Review of Biochemistry, 71*, 307–331.
Ryan, L., Nadel, L., Keil, K., Putnam, K., Schnyer, D., Trouard, T., et al. (2001). Hippocampal complex and retrieval of recent and very remote autobiographical memories: Evidence from functional magnetic resonance imaging in neurologically intact people. *Hippocampus, 11*, 707–714.
Sacks, O. (1985). *The twins. In the man who mistook his wife for a hat.* London: Duckworth.
Sacks, O. (2007). *Musicophilia: Tales of music and the brain.* New York: Knopf Publishing Group.

Sadock, B. J., & Sadock, V. A. (2008). *Kaplan and Sadock's concise textbook of clinical psychiatry* (3rd. ed). Philadelphia: Lippincott, Williams, and Wilkins.

Saenz, D. A., Goldin, A. P., Minces, L., Chianelli, M., Sarmiento, M. I., & Rosenstein, R. E. (2004). Effect of melatonin on the retinal glutamate/glutamine cycle in the golden hamster retina. *FASEB Journal, 18*, 1912–1913.

Salley, R. D. (1988). Subpersonalities with dreaming functions in a patient with multiple personalities. *Journal of Nervous and Mental Disease, 176*, 112–115.

Sammer, G. (1996). Working-memory load and dimensional complexity of the EEG. *International Journal of Psychophysiology, 24*, 173–182.

Sannita, W. (2000). Stimulus-specific oscillatory responses of the brain: A time/frequency related coding process. *Clinical Neurophysiology, 11*, 565–583.

Saper, C. B., Lu, J., Chou, T. C., & Gooley, J. (2005). The hypothalamic integrator for circadian rhythms. *Trends in Neurosciences, 28*, 152–157.

Sapolsky, R. M., Romero, L. M., & Munck, A. U. (2000). How do glucocorticoids influence stress responses? Integrating permissive, suppressive, stimulatory, and preparative actions. *Endocrine Reviews, 21*, 55–89.

Sar, V., & Ross, C. (2006). Dissociative as a confounding factor in psychiatric research. *Psychiatric Clinics of North America, 29*, 129–144.

Sarkar, S. (2005). Einstein's universe: The challenge of dark energy. *Current Science, 89*, 2120–2128.

Sarrieau, A., Dussaillant, M., & Moguilewsky, M. (1988). Autoradiographic localization of glucocorticosteroid binding sites in rat brain after in vivo injection of [3H]RU 28362. *Neuroscience Letters, 92*, 14–20.

Sauseng, P., & Klimesch, W. (2008). What does phase information of oscillatory brain activity tell us about cognitive processes? *Neuroscience and Biobehavioral Reviews, 32*, 1001–1013.

Saver, J. L., & Rabin, J. (1997). The neural substrates of religious experience. *Journal of Neuropsychiatry and Clinical Neurosciences, 9*, 498–510.

Savitz, J., Solms, M., & Ramesar, R. (2006). The molecular genetics of cognition: Dopamine, COMT and BDNF. Genes, Brain and. *Behavior, 5*, 311–328.

Saxe, G. N., Chawla, N., & van der Kolk, B. (2002). Self-destructive behavior in patients with dissociative disorders. *Suicide and Life-Threatening Behavior, 32*, 313–319.

Schein, S. J., & Desimone, R. (1990). Spectral properties of V4 neurons in the macaque. *Journal of Neuroscience, 10*, 3369–3389.

Schenk, L., & Bear, D. (1981). Multiple personalityand related dissociative phenomena in patients with temporal lobeepilepsy. *American Journal of Psychiatry, 138*, 1311–1316.

Schlosshauer, M. (2005). Decoherence, the measurement problem, and interpretations of quantum mechanics. *Reviews of Modern Physics, 76*, 1267–1305.

Schmahmann, J., & Sherman, J. C. (1998). The cerebellar cognitive affective syndrome. *Brain, 121*, 561–579.

Schmid, G. B. (1991). Chaos theory and schizophrenia: Elementary aspects. *Psychopathology, 24*, 185–198.

Schmitz, E. B., Robertson, M. M., & Trimble, M. R. (1999). Depression and schizophreniain epilepsy: Social and biological risk factors. *Epilepsy Research, 35*, 59–68.

Schore, A. N. (1997). A century after Freud's project: Is a rapprochement between psychoanalysis and neurobiology at hand? *Journal of American Psychoanalytic Association, 45*, 807–840.

Schore, A. N. (2001). The effects of early relational trauma on right brain development, affect regulation and infant mental health. *Infant Mental Health Journal, 22*, 201–269.

Schore, A. N. (2002). Dysregulation of the right brain: A fundamental mechanism of traumatic attachment and the psychopathogenesis of posttraumatic stress disorder. *Australian and New Zealand Journal of Psychiatry, 36*, 9–30.

Schrödinger, E. (1944). *What is life? The physical aspect of the living cell.* Cambridge: Cambridge University Press.

Schulte-Herbruggen, O., Nassenstein, C., Lommatzsch, M., Quarcoo, D., Renz, H., & Braun, A. (2005). Tumor necrosis factor-alpha and interleukin-6 regulate secretion of brain-derived neurotrophic factor in human monocytes. *Journal of Neuroimmunology, 160*, 204–209.

Schutter, D. L. J. G., & van Honk, J. (2006). An electrophysiological link between the cerebellum, cognition and emotion: Frontal theta EEG activity to single-pulse cerebellar TMS. *NeuroImage, 33*, 1227–1231.

Schwartz, J. M., Stapp, H. P., & Beauregard, M. (2005). Quantum physics in neuroscience and psychology: A neurophysical model of mind-brain interaction. *Philosophical Transactions of the Royal Society B: Biological Sciences, 360*, 1309–1327.

Scott, A. (Ed.). (2005). *Encyclopedia of nonlinear science*. New York: Routledge, Taylor and Francis Group.

Segal, Z. V., Williams, J. M., Teasdale, J. D., & Gemar, M. (1996). A cognitive science perspective on kindling and episode sensitization in recurrent affective disorder. *Psychological Medicine, 26*, 371–380.

Sells, R. L., & Weidner, R. T. (1980). *Elementary modern physics*. Boston: Allyn and Bacon.

Servit, Z., & Musil, F. (1981). Prophylactic treatment of posttraumatic epilepsy: Results of a long-term follow-up in Czechoslovakia. *Epilepsia, 22*, 315–320.

Seth, A. K., Izhikevich, E., Reeke, G. N., & Edelman, G. M. (2006). Theories and measures of consciousness: An extended framework. *Proceedings of the National Academy of Sciences of the United States of America, 103*, 10799–10804.

Shakow, D. (1980). Kent-Rosanoff association and its implications for segmental set theory. *Schizophrenia Bulletin, 6*, 676–685.

Shannon, C. E. (1948). A mathematical theory of communication. *Bell System Technical Journal, 27*, 379-423 and 623–656.

Shannon, C. E., & Weaver, W. (1963). *A mathematical theory of communication*. Champaign: University of Illinois Press.

Sherry, D. F., & Schacter, D. L. (1987). The evolution of multiple memory systems. *Psychological Review, 94*, 439–454.

Sherry, D. F., Jacobs, L. F., & Gaulin, S. J. C. (1992). Spatial memory and adaptive specialization of the hippocampus. *Trends in Neurosciences, 15*, 298–303.

Shettleworth, S. J. (2009). *Cognition, evolution, and behavior*. New York: Oxford University Press.

Shevrin, H. (2001). Event-related markers of unconscious processes. *International Journal of Psychophysiology, 42*, 209–218.

Shibata, S., Cassone, V. M., & Moore, R. Y. (1989). Effects of melatonin on neuronal activity in the rat suprachiasmatic nucleus in vitro. *Neuroscience Letters, 97*, 140–144.

Sieniutycz, S., & Salamon, P. (1990). *Finite-time thermodynamics and thermoeconomics*. New York: Taylor and Francis.

Sierra, M., & Berrios, G. E. (1998). Depersonalization: Neurobiological perspectives. *Biological Psychiatry, 44*, 898–908.

Silberman, E., Post, R., Nurenberger, J., Theodore, W., & Boulenger, J. (1985). Transient sensory, cognitive, and affective phenomena in affective illness: A comparison with complex partial epilepsy. *British Journal of Psychiatry, 146*, 81–89.

Silva, A. J., Kogan, J. H., Frankland, P. W., & Kida, S. (1998). CREB and memory. *Annual Review of Neuroscience, 21*, 127–148.

Simonneaux, V., & Ribelayga, C. (2003). Generation of the melatonin endocrine message in mammals: A review of the complex regulation of melatonin synthesis by norepinephrine peptides, and other pineal transmitters. *Pharmacological Reviews, 55*, 325–395.

Singer, W. (1993). Synchronization of cortical activity and its putative role in information procesing and learning. *Annual Review of Physiology, 55*, 349–374.

Singer, W. (2001). Consciousness and the binding problem. *Annals of the New York Academy of Sciences, 929*, 123–146.

Singer, W., & Gray, C. M. (1995). Visual feature integration and the temporal correlation hypothesis. *Annual Review of Neuroscience, 18*, 555–586.

Skaper, S. D., Ancona, B., Facci, L., Franceschini, D., & Giusti, P. (1998). Melatonin prevents the delayed death of hippocampal neurons induced by enhanced excitatory neurotransmission and the nitridergic pathway. *FASEB Journal, 12*, 725–731.

Skarda, Ch. A., & Freeman, W. J. (1987). How brains make chaos in order to make sense of the world. *Behavioural and Brain Sciences, 10*, 161–195.

Smith, C. U. M. (1998). Descartes' pineal neuropsychology. *Brain and Cognition, 36*, 57–72.

Smith, P. F., & Darlington, C. L. (1996). The development of psychosisin epilepsy: A reexamination of the kindlinghypothesis. *Behavioural Brain Research, 75*, 59–66.

Smolin, L. (2004). Atoms of space and time. *Scientific American, 290*, 66–75.

Snyder, A. (2009). Explaining and inducing savant skills: Privileged access to lower level, less-processed information. *Philosophical Transactions of the Royal Society B: Biological Sciences, 364*, 1399–1405.

Snyder, A. W., & Mitchell, D. J. (1999). Is integer arithmetic fundamental to mental processing? The mind's secret arithmetic. *Proceedings of the Royal Society B: Biological Sciences, 266*, 587–592.

Snyder, A. W., Mulcahy, E., Taylor, J. L., Mitchell, D. J., Sachdev, P., & Gandevia, S. C. (2003). Savant-like skills exposed in normal people by suppressing the left fronto-temporal lobe. *Journal of Integrative Neuroscience, 2*, 149–158.

Solms, M. (1997). What is consciousness? *Journal of the American Psychoanalytic Association, 45*(3), 681–703.

Solms, M. (2004). Freud returns. *Scientific American, 290*(5), 82–88.

Solms, M., & Saling, M. (1986). On psychoanalysis and neuroscience: Freud's attitude to the localizationist tradition. *International Journal of Psychoanalysis, 67*(4), 397–416.

Solms, M., & Turnbull, O. H. (2011). What is neuropsychoanalysis? *Neuropsychoanalysis, 13*(2), 133–146.

Soule, J., Messaoudi, E., & Bramham, C. R. (2006). Brain-derived neurotrophic factor and control of synaptic consolidation in the adult brain. *Biochemical Society Transactions, 34*, 600–604.

Spence, S., Shapiro, D., & Zaidel, E. (1996). The role of the right hemisphere in the physiological and cognitive components of emotional processing. *Psychophysiology, 33*, 112–122.

Sperry, R. W. (1968). Hemisphere deconnection and unity in conscious awareness. *American Psychologist, 23*, 723–733.

Spiegel, D. (1991). Neurophysiological correlates of hypnosisand dissociation. *Journal of Neuropsychiatry and Clinical Neurosciences, 3*, 440–445.

Spiegel, D. (1997). Trauma, dissociation, and memory. *Annals of the New York Acadademy of Sciences, 821*, 225–237.

Spiegel, D., & Cardena, E. (1991). Disintegrated experience: The dissociative disorders revisited. *Journal of Abnormal Psychology, 100*, 366–376.

Spigelman, I., Li, Z., Banerjee, P. K., Mihalek, R. M., Homanics, G. E., & Olsen, R. W. (2002). Behavior and physiology of mice lacking the GABAA-receptor delta subunit. *Epilepsia, 43*, 3–8.

Spitz, H. (1995). Calendar calculating, idiots savants, and the smart unconscious. *New Ideas in Psychology, 13*, 167–192.

Spitzer, C., Willert, C., Grabe, H., Rizos, T., Moller, B., & Freyberger, H. J. (2004). Dissociation, hemispheric asymmetry, and dysfunctionof hemispheric interaction: A transcranial magnetic stimulation approach. *Journal of Neuropsychiatry and Clinical Neurosciences, 16*, 163–169.

Sporns, O., Tononi, G., & Edelman, G. M. (2000). Connectivity and complexity: The relationship between neuroanatomy and brain dynamics. *Neural Networks, 13*, 909–922.

Sporns, O., Tononi, G., & Edelman, G. M. (2002). Theoretical neuroanatomy and the connectivity of the cerebral cortex. *Behavioural Brain Research, 135*, 69–74.

Sramka, M., Sedlak, P., & Nadvornik, P. (1977). Observation of kindling phenomenon in treatment of pain by stimulation in thalamus. In W Sweet (Ed.), *Neurosurgical treatment in psychiatry* (pp. 651–654). Baltimore: University Park Press.

Stam, C. J. (2005). Nonlinear dynamical analysis of EEG and MEG: Review of an emerging field. *Clinical Neurophysiology, 116*, 2266–2301.

Stam, C. J., van Woerkom, T. C., & Pritchard, W. S. (1996). Use of non-linear EEG measures to characterize EEG changes during mental activity. *Electroencephalography and Clinical Neurophysiology, 99*, 214–224.

Stapp, H. P. (2001). Quantum theory and the role of mind in nature. *Foundations of Physics, 31*, 1465–1499.

Stapp, H. P. (2004). Quantum leaps in philosophy of mind: Reply to Bourget's critique. *Journal of Consciousness Studies, 11*, 43–49.

Stapp, H. P. (2005). Quantum interactive dualism: An alternative to materialism. *Journal of Consciousness Studies, 12*, 43–58.

Stehle, J., Vanecek, J., & Vollrath, L. (1989). Effects of melatonin on spontaneous electrical activity of neurons in rat suprachiasmatic nuclei: An in vitro iontophoretic study. *Journal of Neural Transmission, 78*, 173–177.

Stern, E. J., & Riegel, K. F. (1970). Comparisons of the restricted association of chronic schizophrenic and normal control subjects. *Journal of Abnormal Psychology, 75*, 164–171.

Steuer, R., Ebeling, W., Bengner, T., Dehnicke, C., Hättig, H., & Meencke, H. (2004). Entropy and complexity analysis of intracranially recorded EEG. *International Journal of Bifurcation and Chaos in Applied Sciences and Engineering, 14*, 815–823.

Stevens, J. R. (1999). Epilepsy, schizophreniaand the extended amygdala. *Annals of the New York Academy of Sciences, 156*, 548–561.

Stevens, J. R. (2002). Schizophrenia: Reproductive hormones and the brain. *American Journal of Psychiatry, 159*, 713–719.

Stross, L., & Shevrin, H. (1962). Differences in thought organization between hypnosis and the waking state: An experimental approach. *Bulletin of the Meninger Clinic, 26*, 237–247.

Stross, L., & Shevrin, H. (1968). Thought organization in hypnosis and the waking state. *Journal of Nervous and Mental Disease, 147*, 272–288.

Stross, L., & Shevrin, H. (1969). Hypnosis as a method for investigating unconscious thought processes. *Journal of the American Psychoanalytic Association, 17*, 100–135.

Sudarshan, E. C. G., & Misra, B. (1977). The Zeno's paradox in quantum theory. *Journal of Mathematical Physics, 18*, 756–763.

Sullivan, R. M., & Gratton, A. (1999a). Lateralized effects of medial prefrontal cortex lesions on neuroendocrine and autonomic stress responses in rats. *Journal of Neuroscience, 19*, 2834–2840.

Sullivan, R. M., & Gratton, A. (1999b). Medial prefrontal cortex, laterality, and stress. *Journal of Neuroscience, 19*, 2834–2840.

Sullivan, R. M., & Gratton, A. (2002). Prefrontal cortical regulation of hypothalamic-pituitary-adrenal function in the rat and implications for psychopathology: Side matters. *Psychoneuroendocrinology, 27*, 99–114.

Svetlak, M., Bob, P., Cernik, M., & Kukleta, M. (2010). Electrodermal complexity during the stroop colour word test. *Autonomic Neuroscience: Basic and Clinical, 152*, 101–107.

Szilard, L. (1929/1964). On the decrease of entropy in a thermodynamic system by the intervention of intelligent beings [In German published in 1929, Über die Entropieverminderung in einem thermodynamischen System bei Eingriffen intelligenter Wesen, Zeitschrift für Physik, 53, 840–856]. In English published in 1964. *Behavioral Science, 9*, 301–310.

Talvitie, V., & Ihanus, J. (2011). On neuropsychoanalytic metaphysics. *International Journal of Psychoanalysis, 92*(6), 1583–1601.

Tammet, D. (2006). *Born on a blue day*. New York: Free Press.

Tang, Y. Y., Lu, Q., Geng, X., Stein, E. A., Yang, Y., & Posner, M. I. (2010). Short-term meditation induces white matter changes in the anterior cingulate. *Proceedings of the National Academy of Sciences of the United States of America, 107*, 15649–15652.

Tang, Y. Y., Lu, Q., Fan, M., Yang, Y., & Posner, M. I. (2012). Mechanisms of white matter changes induced by meditation. *Proceedings of the National Academy of Sciences of the United States of America, 109*, 10570–10574.

Tegmark, M. (2000). The importance of quantum decoherence in brain processes. *Physical Review E, 61*, 4194–4206.

Teicher, M. H., Glod, C. A., Surrey, J., & Swett, C. Jr. (1993). Early childhood abuse and limbic system ratings in adult psychiatric outpatients. *Journal Neuropsychiatry and Clinical Neurosciences, 5*, 301–306.

Teicher, M. H., Andersen, S. L., Polcari, A., Anderson, C. M., Navalta, C. P., & Kim, D. M. (2003). The neurobiological consequences of early stress and childhood maltreatment. *Neuroscience and Biobehavioral Reviews, 27*, 33–44.

Teicher, M. H., Tomoda, A., & Andersen, S. E. (2006). Neurobiological consequences of early stress and childhood maltreatment: Are results from human and animal studies comparable? *Annals of the New York Academy of Sciences, 1071*, 313–323.

Teicher, M. H., Samson, J. A., Polcari, A., & McGreenery, C. E. (2006b). Sticks, stones, and hurtful words: Relative effects of various forms of childhood maltreatment. *American Journal of Psychiatry, 163*, 993–1000.

Thomas, K., & Davies, A. (2005). Neurotrophins: A ticket to ride for BDNF. *Current Biology, 15*, 262–264.

Thorne, K. S. (1994). *Black holes and time warps: Einstein's outrageous legacy*. New York: W.W. Norton and Company.

Tiihonen, J., Wahlbeck, K., & Kiviniemi, V. (2009). The efficacy of lamotrigine in clozapine-resistant schizophrenia: A systematic review and meta-analysis. *Schizophrenia Research, 109*, 10–14.

Tirsch, W. S., Stude, Ph., Scherb, H., & Keidel, M. (2004). Temporal order of nonlinear dynamics in human brain. *Brain Research Reviews, 45*, 79–95.

Tomberg, C. (1999). Focal enhancement of chaotic strange attractor dimension in the left semantic (Wernicke) human cortex during reading without concomitant change in vigilance level. *Neuroscience Letters, 263*, 177–180.

Tong, F. (2003). Out-of-body experiences: From Penfield to present. *Trends in Cognitive Sciences, 7*, 104–106.

Tononi G. (2005). Consciousness, information integration, and the brain. *Progress in Brain Research, 150*, 109–126.

Tononi, G., & Edelman, G. M. (2000). Schizophrenia and the mechanisms of conscious integration. *Brain Research Reviews, 31*, 391–400.

Tononi, G., Sporns, O., & Edelman, G. M. (1992). Reentry and the problem of integrating multiple cortical areas: Simulation of dynamic integration in the visual system. *Cerebral Cortex, 2*, 310–335.

Tononi, G., Sporns, O., & Edelman, G. M. (1996). A complexity measure for selective matching of signals by the brain. *Proceedings of the National Academy of Sciences of the United States of America, 93*, 3422–3427.

Tononi, G., Edelman, G. M., & Sporns, O. (1998). Complexity and coherency: Integrating information in the brain. *Trends in Cognitive Sciences, 2*, 474–484.

Tononi, G., Srinivasan, R., Russell, D. P., & Edelman, G. M. (1998a). Investigating neural correlates of conscious perception by frequency-tagged neuromagnetic responses. *Proceedings of the National Academy of Sciences of the United States of America, 95*, 3198–3203.

Tononi, G., McIntosh, A. R., Russell, D. P., & Edelman, G. M. (1998b). Functional clustering: Identifying strongly interactive brain regions in neuroimaging data. *NeuroImage, 7*, 133–149.

Torres-Farfan, C., Richter, H. G., Rojas-García, P., Vergara, M., Forcelledo, M. L., Valladares, L. E., Torrealba, F., Valenzuela, G. J., & Serón-Ferré, M. (2003). mt1 melatonin receptor in the primate adrenal gland: Inhibition of adrenocorticotropin-stimulated cortisol production by melatonin. *Journal of Clinical Endocrinology and Metabolism, 88*, 450–458.

Treffert, D. (1988). The idiot savant: A review of the syndrome. *American Journal of Psychiatry, 145*, 563–572.

Treffert, D. A. (1989). *Extraordinary people: Understanding. "idiot savants."* New York: Harper and Row.

Treffert, D. A. (1999). The savant syndrome and autistic disorder. *CNS Spectrums, 4*, 57–60.

Treffert, D. A. (2002). Savant.com: The internet as a medical education tool. *Wisconsin Medical Journal, 101*, 14–16.

Treffert, D. A. (2005). The savant syndrome in autistic disorder. In M. F. Casanova (Ed.), Recent developments in autism research (pp. 27–55, ch. 2). New York: Nova Science Publishers, Inc

Treffert, D. A. (2009). The savant syndrome: An extraordinary condition. A synopsis: Past, present, future. *Philosophical Transactions of the Royal Society B: Biological Sciences, 364*, 1351–1357.

Treue, S., & Andersen, R. A. (1996). Neural responses to velocity gradients in macaque cortical area MT. *Visual Neuroscience, 13*, 797–804.

Trief, P. M. (1996). Childhood abuse and chronic pain. observations from clinical practice. *Journal of Interpersonal Violence, 11*, 599–608.

Trimble, M. R. (1996). Anticonvulsant-induced psychiatric disorders. The role of forced normalization. *Drug Safety, 15*, 159–166.

Tsakiris, M., Haggard, P., Franck, N., Mainy, N., & Sirigu A. (2005). A specific role for efferent information in self-recognition. *Cognition, 96*, 215–231.

Tuchman, R., Cuccaro, M., & Alessandri, M. (2010). Autism and epilepsy: Historical perspective. *Brain and Development, 32*, 709–718.

Turing, A. M. (1952). The chemical basis of morphogenesis. *Philosophical Transactions of the Royal Society of London, 237*(641), 37–72.

Tyson, J. J. (1979). Oscillations, bistability and echo waves in models of the Belousov-Zhabotinskii reaction. *Annals of the New York Academy of Sciences, 316*, 279–295.

Uhlhaas, P. J., & Singer, W. (2010). Abnormal neural oscillations and synchrony in schizophrenia. *Nature Reviews Neuroscience, 11*, 100–113.

Uhlhaas, P. J., Haenschel, C., Nikolić, D., & Singer, W. (2008). The role of oscillations and synchrony in cortical networks and their putative relevance for the pathophysiology of schizophrenia. *Schizophrenia Bulletin, 34*, 927–943.

Uysal, N., Ozdemir, D., Dayi, A., Yalaz, G., Baltaci, A. K., & Bediz, C. S. (2005). Effects of maternal deprivation on melatonin production and cognition in adolescent male and female rats. *Neuroendocrinology Letters, 26*, 555–560.

Vaitl, D., Lipp, O., Bauer, U., Schüler, G., Stark, R., Zimmerman, M., & Kirsch, P. (2002). Latent inhibition in schizophrenia: Pavlovian conditioning of autonomic responses. *Schizophrenia Research, 55*, 147–158.

Valentine, E. R. (1989). Neural nets: From Hartley and Hebb to Hinton. *Journal of Mathematical Psychology, 33*, 348–357.

van Drongelen, W., Lee, H. C., Hereld, M., Chen, Z., Elsen, F. P., & Stevens, R. L. (2005). Emergent epileptiform activity in neural networks with weak excitatory synapses. *IEEE Transactions on Neural Systems and Rehabilitation Engineering, 13*, 236–241.

Van Os, J., Rutten, B. P. F., & Poulton, R. (2008). Gene-environment interactions in schizophrenia: Review of epidemiological findings and future directions. *Schizophrenia Bulletin, 34*, 1066–1082.

Van Putten, M. J. A. M., & Stam, C. J. (2001). Is the EEG really "chaotic" in hypsarrhythmia. *IEEE Engineering in Medicine and Biology Magazine, 20*, 72–79.

Van Winkel, R., Stefanis, N. C., & Myin-Germeys, I. (2008). Psychosocial stress and psychosis. A review of the neurobiological mechanisms and the evidence for gene-stress interaction. *Schizophrenia Bulletin, 34*, 1095–1105.

van de Grind, W. (2002). Physical, neural, and mental timing. *Consciousness and Cognition, 11*, 241–264.

van den Top, M., Buijs, R. M., Ruijter, J. M., Delagrange, P., Spanswick, D., & Hermes, M. L. (2001). Melatonin generates an outward potassium current in rat suprachiasmatic nucleus neurones in vitro independent of their circadian rhythm. *Neuroscience, 107*, 99–108.

Van der Hart, O., & Brown, P. (1992). Abreaction re-evaluated. *Dissociation, 5*, 127–140.

Van der Hart, O., & Friedman, B. (1989). A reader's guide to Pierre Janeton dissociation: A neglected intelectual heritage. *Dissociation, 2*, 3–16.

van der Hart, O., Nijenhuis, E. R., & Steele, K. (2005). Dissociation: An insufficiently recognized major feature of complex posttraumatic stress disorder. *Journal of Traumatic Stress, 18*, 413–423.

van der Kolk, B. A., & van der Hart, O. (1989). Pierre Janet and the breakdown of adaptation of adaptation in psychological trauma. *American Journal of Psychiatry, 146*, 1530–1540.

van der Velde, F., & de Kamps, M. (2006). Neural blackboard architectures of combinatorial structures in cognition. *Behavioral and Brain Sciences, 29*, 1–72.

Varela, F., & Thompson, E. (2003). Neural synchrony and the unity of mind: A neurophenomenological perspective. In A. Cleeremans (Ed.), *The unity of consciousness: Binding, integration, and dissociation* (pp. 266–287). Oxford: Oxford University Press.

Varela, F. J., Lachaux, J. P., Rodriguez, E., & Martinerie, J. (2001). The brainweb: Phase synchronization and large-scale integration. *Nature Reviews Neuroscience, 2*, 229–239.

Varney, N. R., Garvey, M. J., Cook, B. L., Campbell, D. A., & Roberts, R. J. (1993). Identification of treatment-resistant depressives who respond favorably to carbamazepine. *Annals of Clinical Psychiatry, 5*, 117–122.

Velazquez, J. L. P., Cortez, M. A., Snead, O. C. III, & Wennberg, R. (2003). Dynamical regimes underlying epileptiformevents: Role of instabilities and bifurcations in brainactivity. *Physica D, 186*, 205–220.

Velik, R. (2010). From single neuron-firing to consciousness-towards the true solution of the binding problem. *Neuroscience and Biobehavioral Reviews, 34*, 993–1001.

Vermetten, E., & Douglas, B. J. (2004). Functional brain imaging and the induction of traumatic recall: A cross-correlational review between neuroimaging and hypnosis. *International Journal of Clinical and Experimental Hypnosis, 52*, 280–312.

Vichyn, B. (2005). Project for a scientific psychology. In A. De Mijolla (Ed.), *International dictionary of psychoanalysis*. Detroit: Macmillan Reference.

Vignal, J., Maillard, L., McGonigal, A., & Chauvel, P. (2007). The dreamy state: Hallucinations of autobiographic memory evoked by temporal lobe stimulations and seizures. *Brain, 130*(1), 88–99.

Vigo, D. V., & Baldessarini, R. J. (2009). Anticonvulsants in the treatment of major depressive disorder: An overview. *Harvard Review of Psychiatry, 17*, 231–241.

Villemure, Ch., & Bushnell, M. C. (2002). Cognitive modulation of pain: How do attention and emotion influence pain processing? *Pain, 95*, 195–199.

Vinogradov, S., Kirkland, J., Poole, J. H., Drexler, M., Ober, B. A., Shenaut, G. K. (2002). Both processing speed and semantic memory organization predict verbal fluency in schizophrenia. *Schizophrenia Research, 8*, 171–181.

Viscott, D. S. (1970). A musical idiot savant. *Psychiatry, 33*, 494–515.

Vivona, J. M. (2009). Leaping from brain to mind: A critique of mirror neuron explanations of countertransference. *Journal of the American Psychoanalytic Association, 57*, 525–550.

Volkenstein, M. V. (2009). *Entropy and information*. Basel: Birkhauser Verlag AG.

Vollrath, L., & Welker, H. A. (1988). Day-to-day variation in pineal serotonin N-acetyltransferase activity in stressed and nonstressed male Sprague-Dawley rats. *Life Sciences, 42*, 2223–2229.

von Helmholtz, H. (1962). *Treatise on physiological optics*. New York: Dover

von Laue, M. (1950). *History of physics*. New York: Academic Press.

Von Neumann, J. (1955). *Mathematical foundations of quantum mechanics*. Princeton: Princeton University Press.

von der Malsburg, C. (1996). The binding problem of neural networks. In R. Llinas & P. S. Churchland (Eds.), *The mind-brain continuum: Sensory processes* (pp. 131–146). Cambridge: MIT Press.

von der Malsburg, C. (1999). The what and why of binding: The modeler's perspective. *Neuron, 24*, 95–104.

von der Malsburg, C., & Schneider, W. (1986). A neural cocktail-party processor. *Biological Cybernetics, 54*, 29–40.

von Gall, C., Garabette, M. L., Kell, C. A., Frenzel, S., Dehghani, F., Schumm-Draeger, P. M., Weaver, D. R., Korf, H. W., Hastings, M. H., & Stehle, J. H. (2002). Rhythmic gene expression in pituitary depends on heterologous sensitization by the neurohormone melatonin. *Nature Neuroscience, 5*, 234–238.

Wackermann, J., Putz, P., Buchi, S., Strauch, I., & Lehmann, D. (2002). Brain electrical activity and subjective experience during altered states of consciousness: Ganzfeld and hypnagogic states. *International Journal of Psychophysiology, 46*, 123–46.

Wackermann, J., Putz, P., & Allefeld, C. (2008). Ganzfeld-induced hallucinatory experience, its phenomenology and cerebral electrophysiology. *Cortex, 44*, 1364–1378.

Walsh, V., & Cowey, A. (2000). Transcranial magnetic stimulation and cognitive neuroscience. *Nature Reviews Neuroscience, 1*, 73–79.

Wan, Q., Man, H. Y., Liu, F., Braunton, J., Niznik, H. B., Pang, S. F., Brown, G. M., & Wang, Y. T. (1999). Differential modulation of GABAA receptor function by Mel1a and Mel1b receptors. *Nature Neuroscience, 2*, 401–403.

Wang, L. M., Suthana, N. A., Chaudhury, D., Weaver, D. R., & Colwell, C. S. (2005). Melatonin inhibits hippocampal long-term potentiation. *European Journal of Neuroscience, 22*, 2231–2237.

Warembourg, M. (1975). Radioautographic study of the rat brain and pituitary after injection of 3H dexamethasone. *Cell Tissue Research, 161*, 183–191.

Wass, S. (2011). Distortions and disconnections: Disrupted brain connectivity in autism. *Brain and Cognition, 75*, 18–28.

Watkins, H. H. (1993). Ego-state therapy: An overview. *American Journal of Clinical Hypnosis, 35*, 232–240.

Watkins, J. G., & Watkins, H. H. (1979–1980). Ego states and hidden observers. *Journal of Altered States of Consciousness, 5*, 3–18.

Watkins, J. G., & Watkins, H. H. (1990). Dissociation and displacement: Where goes the "ouch?" *American Journal of Clinical Hypnosis, 33*, 1-21.

Weaver, W. (1948). Science and Complexity. *American Scientist, 36*(4), 536–544.

Weingartner, H., Miller, H., & Murphy, D. (1977). Mood-state-dependent retrieval of verbal associations. *Journal of Abnormal Psychology, 86*, 276–284.

Weiskrantz, L. (1998). Consciousness and commentaries. In S. R. Hameroff, A. Kaszriak, & A. C. Scott (Eds.), *Toward a science of consciousness II—The second Tucson discussions and debates* (pp. 371–377). Cambridge: MIT Press.

Welling, H. (1994). Prime number identification in idiot savants: Can they calculate them? *Journal of Autism and Developmental Disorders, 24*, 199–207.

Weng, G., Bhalla, U. S., & Iyengar, R. (1999). Complexity in biological signaling systems. *Science, 284*, 92–96.

Westen, D., & Gabbard, G. O. (2002a). Developments in cognitive neuroscience: I. conflict, compromise, and connectionism. *Journal of the American Psychoanalytic Association, 50*, 53–98.

Westen, D., & Gabbard, G. O. (2002b). Developments in cognitive neuroscience: II. implications for theories of transference. *Journal of the American Psychoanalytic Association, 50*(1), 99–134.

Wheeler, J. A., & Zurek, W. H. (1983). *Quantum theory and measurement*. New Jersey: Princeton University Press.

White, P. A. (1988). The structural representation of information in LTM: A possible explanation for the accomplishments of "idiots savants". *New Ideas in Psychology, 6*, 3–14.

Whiteside, D. T. (Ed.) (1967). *The mathematical papers of Isaac Newton*. (vol. 1, part 7 "The October 1666 Tract on Fluxions"). Cambridge University Press

Wiens, S. (2006). Subliminal emotion perception in brain imaging: Findings, issues, and recommendations. *Progress in Brain Research, 156*, 105–121.

Wilson, T. W., Rojas, D. C., Reite, M. L., Teale, P. D., & Rogers, S. J. (2007). Children and adolescents with autism exhibit reduced MEG steady-state gamma responses. *Biological Psychiatry, 62*, 192–197.

Winter, H., & Irle, E. (2004). Hippocampal volume in adult burn patients with and without post-traumatic stress disorder. *American Journal of Psychiatry, 161*, 2194–2200.

Wittling, W., & Pfluger, M. (1990). Neuroendocrine hemisphere asymmetries: Salivary cortisol secretion during lateralized viewing of emotion-related and neutral films. *Brain and Cognition, 14*, 243–265.

Wolf, P. (1991). Acute behavioural symptomatology at disappearence of epileptiform EEGabnormality. Paradoxical or 'forced' normalization. *Advances in Neurology, 55*, 127–142.

Wolf, P., & Trimble, M. R. (1985). Biological antagonism and epileptic psychosis. *British Journal of Psychiatry, 146*, 272–276.

Wolff, P. H. (1987). *The development of behavioural and emotional states in infancy*. Chicago: University Chicago Press.

Womelsdorf, T., & Fries, P. (2006). Neuronal coherence during selective attentional processing and sensory-motor integration. *Journal of Physiology, Paris, 100*, 182–193.

Womelsdorf, T., & Fries, P. (2007). The role of neuronal synchronization in selective attention. *Current Opinion in Neurobiology, 17*, 154–160.

Wong, P. S., Shevrin, H., & Williams, W. J. (1994). Conscious and Nonconscious processes: An ERP Index of an anticipatory response in a conditioning paradigm using visual masked stimuli. *Psychophysiology, 31*, 87–101.

Wong, P. S., Bernat, E., Snodgrass, M., & Shevrin, H. (2004). Event-related brain correlates of associative learning without awareness. *International Journal of Psychophysiology, 53*, 217–231.

Woolf, N., & Hameroff, S. (2001). A quantum approach to visual consciousness. *Trends in Cognitive Neuroscience, 5*, 472–478.

World Health Organization. (1993). *The ICD-10. Classification of mental and behavioural disorders. Diagnostic criteria for research*. Geneva: World Health Organization.

Wortman, C. B., Loftus, E. F., & Marshall, M. E. (1992). Sensation and Perception. In C. B. Wortman, E. F. Loftus, & M. E. Marshall (Eds.), *Psychology*. New York: Mc Graw-Hill.

Yamada, K, & Nabeshima, T. (2003). Brain-derived neurotrophic factor/TrkB signaling in memory processes. *Journal of Pharmacological Science, 91*, 267–270.

Yates, J. L., & Nasby, W. (1993). Dissociation, affect, and network models of memory: An integrative proposal. *Journal of Traumatic Stress, 6*, 305–326.

Young, R. L., Ridding, M. C., & Morrell, T. L. (2004). Switching skills on by turning off part of the brain. *Neurocase, 10*, 215–222.

Yuii, K., Suzuki, M., & Kurachi, M. (2007). Stress sensitization in schizophrenia. *Annals of the New York Academy of Sciences, 1113*, 276–290

Zaccai, G., Massoulié, J., & David, F. (1998). *From cell to brain: The cytoskeleton, intra- and inter-cellular communication, the central nervous system*. Amsterdam: Elsevier.

Zaikin, A. N., & Zhabotinsky, A. M. (1970). Concentration wave propagation in two-dimensional liquid-phase self-oscillating system, *Nature, 225*, 535–537.

Zakharov, A. F., Capozziello, S., De Paolis, F., Ingrosso, G., & Nucita, A. A. (2009). The role of dark matter and dark energy in cosmological models: Theoretical overview. *Space Science Reviews, 148*, 301–313

Zeise, M.L., & Semm, P. (1985) Melatonin lowers excitability of guinea pig hippocampal neurons in vitro. *Journal of Comparative Physiology, 57*, 23–29.

Zeki, S. (1994). *A vision of the brain*. Oxford: Blackwell Science.

Zeki, S. (2003). The disunity of consciousness. *Trends in Cognitive Neuroscience, 7*, 214–218.

Zeki, S., & Bartels, A. (1999). Toward a theory of visual consciousness. *Consciousness and Cognition, 8*, 225–259.

Zeman, A. (2001). Consciousness. *Brain, 124*(Pt 7):1263–1289.

Zurek, W. H. (1984). Maxwell's Demon, Szilard's engine and quantum measurements, In G. T. Moore & M. O. Scully (Eds), *Frontiers of nonequilibrium statistical physics* (pp. 151–160). New York: Plenum Press.

Index

P. Bob, *The Brain and Conscious Unity,* DOI 10.1007/978-1-4939-2700-5